T0255065

Problem Books in Mathematics

Series editor:

Peter Winkler
Department of Mathematics
Dartmouth College
Hanover, NH 03755
USA

More information about this series at http://www.springer.com/series/714

Volodymyr Brayman · Alexander Kukush

Undergraduate Mathematics Competitions (1995–2016)

Taras Shevchenko National University of Kyiv

Second Edition

Volodymyr Brayman
Department of Mathematical Analysis
Taras Shevchenko National University of Kyiv
Kyiv
Ukraine

Alexander Kukush
Department of Mathematical Analysis
Taras Shevchenko National University of Kyiv
Kyiv
Ukraine

ISSN 0941-3502 ISSN 2197-8506 (electronic)
Problem Books in Mathematics
ISBN 978-3-319-86451-8 ISBN 978-3-319-58673-1 (eBook)
DOI 10.1007/978-3-319-58673-1

Printed on acid-free paper

This Springer imprint is published by Springer Nature
The registered company is Springer International Publishing AG
The registered company address is: Gewerbestrasse 11, 6330 Cham, Switzerland

To our Teachers
Anatoliy Dorogovtsev
and Myhailo Yadrenko

Foreword

The book contains the problems from the last 22 years of the Undergraduate Mathematics Competition at the Mechanics and Mathematics Faculty of Taras Shevchenko National University of Kyiv. The competition has had a long tradition going back to the 1970s. It eventually became a popular competition open to students from other colleges and universities. In the last couple of decades the winners of the competition have participated in the International Mathematical Competition for university students. The Undergraduate Mathematics Competition has provided a good training and selection venue from of the Taras Shevchenko University for composing a successful team for the IMC. The author of this Foreword also participated in the Competition when he was a student. It was a useful and interesting experience, which was very much appreciated.

The problems in this collection are all original, and were mostly written by mathematicians from Kyiv University, but some were also written by mathematicians of other institutions in different countries. They cover a wide variety of areas of mathematics: calculus, algebra, combinatorics, functional analysis, etc. I would especially note that there are many interesting problems in probability theory. Problems are non-standard and solving them requires ingenuity and a deep understanding of the material. The book also contains the original solutions to the problems, many of which are very elegant and interesting to read. This is the second edition of the collection (the first was published in Ukrainian). I am sure that this book will be useful to students and professors as a source of interesting problems for competitions, for training, or even as a collection of harder problems for university courses. The authors of the book, Volodymyr Brayman and Alexander Kukush, are longtime organizers of the Competition. They are professors at the Department of Mathematical Analysis of the Mechanics and Mathematics Faculty of Taras Shevchenko National University of Kyiv, and are active in popularizing mathematics in Ukraine through mathematical olympiads, journals, and books.

They both were winners of the Undergraduate Mathematics Competition. A. Kukush, in particular, was a winner of the Competition in its early years (in 1977 and 1978).

April 2017

Volodymyr Nekrashevych
Professor of Mathematics at Texas
A&M University, College Station, TX, USA

Preface

The Mathematics Olympiad for students of the Mechanics and Mathematics Faculty has been organized at Taras Shevchenko National University of Kyiv since 1974. After a while the competition opened up to qualified students from any higher school of Kyiv and beginning in 2004, it became a nice tradition to invite the strongest mathematics students of leading Kyiv high schools to participate. Since then representatives of Ukrainian Physics and Mathematics Lyceum, Liceum No. 171 "Leader", Liceum "Naukova Zmina", Liceum No. 208, and Rusanivky Liceum have repeatedly become prize winners of the Olympiad.

Most of the Olympiad winners are students of the Mechanics and Mathematics Faculty, but students from the following departments or institutions have also performed successfully: Institute of Physics and Technology and Institute of Applied System Analysis of National Technical University of Ukraine "Igor Sikorsky Kyiv Polytechnic Institute", Faculty of Cybernetics and Faculty of Physics of Taras Shevchenko National University of Kyiv, National Pedagogical Dragomanov University, and National University of Kyiv-Mohyla Academy.

Results of the Olympiad are taken into account when forming teams of All-Ukrainian students' Mathematics Olympiad, International Mathematics Competition for University Students (IMC) and other student competitions. Materials and results of many mathematics competitions in which Ukrainian students take part can be found on the students' page of this website of Mechanics and Mathematics Faculty http://www.mechmat.univ.kiev.ua.

As a rule, first- and second-year undergraduates and third- and fourth-year undergraduate students compete separately. Along the history of the Olympiad, the number of problems distributed has changed several times. Most recently, the jury of Olympiad composed two sets of problems—one for first- and second-year undergraduates and the second set for senior undergraduate students. Each set contained 7–10 problems. For first-and second-year undergraduates, problems were included for fields such as calculus, algebra, number theory, geometry, and discrete mathematics. Problem sets for third and fourth year undergraduates included additional topics in measure theory, functional analysis, probability theory, complex analysis, differential equations, etc. Solutions to all the problems do not rely on statements out of curriculum of obligatory courses studied at Mechanics and Mathematics Faculty, but the solutions demand creative usage of obtained knowledge. Most of the problems are not technical and admit a short and elegant solution. A few complicated problems, which demand general mathematical culture and remarkable inventiveness, are included in both versions of the assignment, and this helps to compare the results of all the participants.

In 1997–1999 some of the problems were borrowed from Putnam Competitions [1, 3, 4]. Almost all the problems of the last 17 years are original. Their authors are lecturers, Ph.D. students, senior students, and graduating students of the Mechanics and Mathematics Faculty, as well as colleagues from Belgium, Canada, Great Britain, Hungary, and the USA. Since 2003 participants obtain an assignment, where the author's name is indicated beside the corresponding problem.

The competition lasts for 3 hours. Of course, this time interval is not enough to solve all the problems, and therefore, a participant can focus first of all on the problems, which are the most interesting for him/her. Typically, almost all the problems are solved by some of participants; a winner solves more than half of problems, and all who solve at least 2–3 problems become prize winners or get the letter of commendation. The jury of olympiad checks the works and gives a preliminary evaluation. Approximately one week later, an analysis of problems is held, appeal, and winners are awarded.

For many years, until 1995, the jury leader was also the head of Mathematical Analysis Department, Prof. Anatoliy Yakovych Dorogovtsev (1935–2004), a famous expert in mathematical statistics and the theory of stochastic equations. For a long time he led a circle in calculus for first- or second-year undergraduate students (until now such circles work at Faculty of Mechanics and Mathematics and at Institute of Mathematics of the National Academy of Sciences of Ukraine). Anatoliy Yakovych proposed numerous witty problems in calculus, measure theory, and functional analysis. For a few years a jury leader was also the head of the Probability Theory and Mathematical Statistics Department as well as a Corresponding Member of the NAS of Ukraine, Myhailo Yosypovych Yadrenko (1932–2004). Myhailo Yosypovych was an outstanding expert in the theory of random fields and had authored many clever problems in probability theory and discrete mathematics. In particular years, the organizers of Olympiad were a Corresponding Member of the NAS of Ukraine Volodymyr Vladyslavovych Anisimov, lecturers Oleksiy Yuriyovych Konstantinov, Volodymyr Stepanovych Mazorchuk, and Volodymyr Volodymyrovych Nekrashevych. From 1999 until now, the permanent jury leader has also been the head of Mathematical Analysis Department, Prof. Igor Oleksandrovych Shevchuk, a famous expert in approximation theory. Members of jury for the last Olympiads were Andriy Bondarenko, Volodymyr Brayman, Alexander Kukush, Yevgen Makedonskyi, Dmytro Mitin, Oleksiy Nesterenko, Vadym Radchenko, Oleksiy Rudenko, Vitaliy Senin, Sergiy Shklyar, Sergiy Slobodyanyuk, and Yaroslav Zhurba.

There are several famous mathematicians among the former winners of the Olympiad of Mechanics and Mathematics Faculty. In particular, Prof. O.G. Reznikov (1960–2003) used powerful methods of calculus in problems of modern geometry and was a member of London Mathematical Society. In 2016 Dr. M.S. Viazovska was awarded the Salem Prize for a conceptual breakthrough in the sphere packing problem. In 2013 Dr. A.V. Bondarenko was awarded the Vasil Popov International Prize for outstanding achievements in approximation theory. State prizes of Ukraine were awarded: to Prof. A.Ya. Dorogovtsev for a monograph in stochastic analysis; D.Sc. in Physics and Mathematics V.V. Lyubashenko for a cycle of papers in algebra; D.Sc. in Physics and Mathematics O.Yu. Teplinskyi for papers in theory of dynamical systems. Candidate of Sciences in physics and mathematics A.V. Knyazyuk (1960–2013) was a famous teacher of the Kyiv Natural Science Luceum No. 145. We mention also Professors I.M. Burban, O.Yu. Daletskyi, P.I. Etingof, M.V. Kartashov, Yu. G. Kondratyev, K.A. Kopotun, A.G. Kukush, O.M. Kulik, V.S. Mazorchuk, Yu. S. Mishura, V.V. Nekrashevych, A.Yu. Pylypenko, V.M. Radchenko, V.G. Samoylenko, G.M. Shevchenko, and B.L. Tsyagan. We apologize if we have forgotten anybody.

The first part of the book contains all the problems of Olympiads dated 1995–2016. We hope that you will enjoy both self-reliant problem solving and an acquaintance with the solutions presented in the second part of the book. Some problems from earlier Olympiads can be found in the articles [2, 5, 6].

The authors are sincerely grateful to Dmytro Mitin for his long-lived fruitful cooperation, and also to Danylo Radchenko and Oleksandr Tolesnikov for useful discussions.

Kyiv, Ukraine
April 2017

Volodymyr Brayman
Alexander Kukush

Contents

Part I Problems

Part I
Problems

Published sets of examination questions contain (for good reasons) not what was set but what ought to have been set; a year with no correction is rare. One year a question was so impossibly wrong that we substituted a harmless dummy.

John E. Littlewood, "*A Mathematician's Miscellany*"

1995

1. (1-year) Prove that for every $n \in \mathbb{N}$ there exists a unique $t(n) > 0$ such that $(t(n) - 1)\ln t(n) = n$. Calculate $\lim\limits_{n\to\infty} \left(t(n)^{\frac{\ln n}{n}}\right)$.

2. (1-year) Let $\{a_n,\ n \geq 1\} \subset \mathbb{R}$ be a bounded sequence. Define

$$b_n = \frac{1}{n}(a_1 + \ldots + a_n),\ n \geq 1.$$

Assume that the set A of partial limits of $\{a_n,\ n \geq 1\}$ coincides with the set of partial limits of $\{b_n,\ n \geq 1\}$. Prove that A is either a segment or a single point. Prove or disprove the following: if A is either a segment or a single point then A and B coincide.

3. (1-year) Let $f\colon \mathbb{R} \to \mathbb{R}$ have a primitive function F on $\mathbb{R}$ and satisfy $2xF(x) = f(x)$, $x \in \mathbb{R}$. Find f.

4. (1-year) Let $f \in C([0, 1])$. Prove that there exists a number $c \in (0, 1)$ such that $\int\limits_0^c f(x)dx = (1 - c)f(c)$.

5. (1–2-years) A sequence of $m \times m$ real matrices $\{A_n,\ n \geq 0\}$ is defined as follows: $A_0 = A$, $A_{n+1} = A_n^2 - A_n + \frac{3}{4}I$, $n \geq 0$, where A is a positive definite matrix such that $\operatorname{tr}(A) < 1$, and I is the identity matrix. Find $\lim\limits_{n\to\infty} A_n$.

6. (1–2-years) Let $\{x_n,\ n \geq 1\} \subset \mathbb{R}$ be a bounded sequence and a be a real number such that $\lim\limits_{n\to\infty} \frac{1}{n}\sum\limits_{k=1}^{n} x_k^j = a^j$, $j = 1, 2$. Prove that $\lim\limits_{n\to\infty} \frac{1}{n}\sum\limits_{k=1}^{n} \sin x_k = \sin a$.

V. Brayman and A. Kukush, *Undergraduate Mathematics Competitions (1995–2016)*, Problem Books in Mathematics,
DOI 10.1007/978-3-319-58673-1_1

7. (1–4-years) Let F be any quadrangle with area 1 and G be a disc with radius $\frac{1}{\pi}$. For every $n \geq 1$, let $a(n)$ be the maximum number of figures of area $\frac{1}{n}$ similar to F with disjoint interiors, which is possible to pack into G. In a similar way, define $b(n)$ as the maximum number of discs of area $\frac{1}{n}$ with disjoint interiors, which is possible to pack into F. Prove that $\limsup\limits_{n\to\infty} \frac{b(n)}{n} < \lim\limits_{n\to\infty} \frac{a(n)}{n} = 1$.

8. (1–4-years) Find the maximal length of a convex piecewise-smooth contour with diameter d.

9. (2-year) Prove that the equation

$$y'(x) - (2 + \cos x)y(x) = \arctan x, \ x \in \mathbb{R},$$

has a unique bounded on $\mathbb{R}$ solution in the class $C^{(1)}(\mathbb{R})$.

10. (2-year) Find all the solutions to the Cauchy problem

$$\begin{cases} y'(x) = \int_0^x \sin(y(x))du + \cos x, \ x \geq 0, \\ y(0) = 0. \end{cases}$$

11. (3-year) A series $f(z) = \sum\limits_{n=0}^{\infty} c_n z^n$ has a unit radius of convergence, and $c_n = 0$ for $n = km + l$, $m \in \mathbb{N}$, where $k \geq 2$ and $0 \leq l \leq k-1$ are fixed. Prove that f has at least two singular points on the unit circle.

12. (3–4-years) Let $K = \{z \in \mathbb{C} \mid 1 \leq |z| \leq 2\}$. Consider the set W of functions u which are harmonic in K and satisfy $\int\limits_{S_j} \frac{\partial u}{\partial n} ds = 2\pi$, where

$$S_j = \{z \in \mathbb{C} \mid |z| = j\}, \ j = 1, 2,$$

and n is a normal to S_j inside K. Let $u^* \in W$ be such a function that $D(u^*) = \min\limits_{u \in W} D(u)$, where

$$D(u) = \iint\limits_K \left(u'^2_x + u'^2_y\right) dxdy.$$

Prove that u^* is constant on both S_1 and S_2.

13. (3–4-years) Each positive integer is a trap with probability 0.4 independently of other integers. A hare is jumping over positive integers. It starts from 1 and jumps each time to the right at distance 0, 1, or 2 with probability $\frac{1}{3}$ and independently of previous jumps. Prove that the hare will be trapped eventually with probability 1.

14. (4-year) Let H be a Hilbert space and A_n, $n \geq 1$ be continuous linear operators such that for every $x \in H$ it holds $\|A_n x\| \to \infty$, as $n \to \infty$. Prove that for every compact operator K it holds $\|A_n K\| \to \infty$, as $n \to \infty$.

THE PROBLEMS ARE PROPOSED BY A.Ya. Dorogovtsev (1, 4) and A.G. Kukush (5, 6).

1996

1. Let $a, b, c \in \mathbb{C}$. Find $\limsup_{n\to\infty} |a^n + b^n + c^n|^{1/n}$.

2. A function $f \in C([1, +\infty))$ is such that for every $x \geq 1$ there exists a limit

$$\lim_{A\to\infty} \int_A^{Ax} f(u)du =: \varphi(x),$$

$\varphi(2) = 1$, and moreover the function φ is continuous at point $x = 1$. Find $\varphi(x)$.

3. A function $f \in C([0, +\infty))$ is such that

$$f(x) \int_0^x f^2(u)du \to 1, \text{ as } x \to +\infty.$$

Prove that

$$f(x) \sim \left(\frac{1}{3x}\right)^{1/3}, \text{ as } x \to +\infty.$$

4. Find

$$\sup_\lambda \left(\frac{\sum_{k=0}^{n-1} (x_{k+1} - x_k) \sin 2\pi x_k}{\sum_{k=0}^{n-1} (x_{k+1} - x_k)^2} \right),$$

where the supremum is taken over all possible partitions of $[0, 1]$ of the form $\lambda = \{0 = x_0 < x_1 < \ldots < x_{n-1} < x_n = 1\}$, $n \geq 1$.

V. Brayman and A. Kukush, *Undergraduate Mathematics Competitions (1995–2016)*, Problem Books in Mathematics,
DOI 10.1007/978-3-319-58673-1_2

5. Find general form of a function $f(z)$, which is analytic on the upper half-plane except the point $z = i$, and satisfies the following conditions:
◇ the point $z = i$ is a simple pole of $f(z)$;
◇ the function $f(z)$ is continuous and real-valued on the real axis;
◇ $\lim\limits_{\substack{z\to\infty \\ \operatorname{Im} z\geq 0}} f(z) = A$ $(A \in \mathbb{R})$.

6. Let $\mathscr{D}$ be a bounded connected domain with boundary $\partial\mathscr{D}$, and $f(z)$, $F(z)$ be functions analytic on $\overline{\mathscr{D}}$. It is known that $F(z) \neq 0$ and $\operatorname{Im} \frac{f(z)}{F(z)} \neq 0$ for every $z \in \partial\mathscr{D}$. Prove that the functions $F(z)$ and $F(z) + f(z)$ have equal number of zeros in $\mathscr{D}$.

7. A linear operator A on a finite-dimensional space satisfies

$$A^{1996} + A^{998} + 1996I = 0.$$

Prove that A has an eigenbasis. Here I is the unit operator.

8. Let $A_1, A_2, \ldots, A_{n+1}$ be $n \times n$ matrices. Prove that there exist numbers a_1, a_2, $\ldots$, a_{n+1} (not all of them equal 0) such that a matrix

$$a_1 A_1 + \ldots + a_{n+1} A_{n+1}$$

is singular.

9. The trace of a matrix A equals 0. Prove that A can be decomposed into a finite sum of matrices, such that the square of each of them equals to zero matrix.

1997

Problems for 1–4-Years Students

1. Let $1 \le k \le n$. Consider all possible decompositions of n into a sum of two or more positive integer summands. (Two decompositions that differ by order of summands are assumed distinct.) Prove that the summand equal k appears exactly $(n-k+3)2^{n-k-2}$ times in the decompositions.

2. Prove that the field $\mathbb{Q}(x)$ of rational functions contains two subfields F and K such that $[\mathbb{Q}(x) : F] < \infty$ and $[\mathbb{Q}(x) : K] < \infty$, but $[\mathbb{Q}(x) : (F \bigcap K)] = \infty$.

3. Let a matrix $A \in M_n(\mathbb{C})$ have a unique eigenvalue a. Prove that A commutes only with polynomials of A if and only if $\mathrm{rk}(A - aI) = n - 1$. Here I is the identity matrix.

4. Solve an equation

$$2^x = \frac{2}{3}x^2 + \frac{1}{3}x + 1.$$

5. Find a limit

$$\lim_{n\to\infty} \left(\int_0^1 e^{x^2/n} dx \right)^n .$$

6. Let $a \in \mathbb{R}^m$ be a column vector and I be the identity matrix of size m. Simplify

$$\left(1 - a^{\mathrm{T}} \left(I + aa^{\mathrm{T}}\right) a\right)^{-1} .$$

7. Find the global maximum of a function $f(x) = e^{\sin x} + e^{\cos x}$, $x \in \mathbb{R}$.

V. Brayman and A. Kukush, *Undergraduate Mathematics Competitions (1995–2016)*, Problem Books in Mathematics,
DOI 10.1007/978-3-319-58673-1_3

8. Let f be a positive nonincreasing function on $[1, +\infty)$ such that

$$\int_1^{+\infty} x f(x) dx < \infty.$$

Prove convergence of an integral

$$\int_1^{+\infty} \frac{f(x)}{|\sin x|^{1-\frac{1}{x}}} dx.$$

9. See William Lowell Putnam Mathematical Competition, 1961, Morning Session, Problem 3.

10. See William Lowell Putnam Mathematical Competition, 1962, Morning Session, Problem 4.

11. Non-constant complex polynomials P and Q have the same set of roots (possibly of different multiplicities), and the same is true for the polynomials $P + 1$ and $Q + 1$. Prove that $P \equiv Q$.

THE PROBLEMS ARE PROPOSED BY O. G. Ganyushkin (1–3) and A. G. Kukush (4–8).

1998

Problems for 1–4-Years Students

1. See William Lowell Putnam Mathematical Competition, 1996, Problem B1.

2. See William Lowell Putnam Mathematical Competition, 1989, Problem A4.

3. See William Lowell Putnam Mathematical Competition, 1997, Problem B6.

4. Let $q \in \mathbb{C}$, $q \neq 1$. Prove that for every nonsingular matrix $A \in M_n(\mathbb{C})$ there exists a nonsingular matrix $B \in M_n(\mathbb{C})$ such that $AB - qBA = I$.

5. See William Lowell Putnam Mathematical Competition, 1992, Problem B6.

6. See William Lowell Putnam Mathematical Competition, 1989, Problem A6.

7. See William Lowell Putnam Mathematical Competition, 1997, Problem B2.

8. Does there exist a function $f \in C(\mathbb{R})$ such that for every real number x it holds

$$\int_0^1 f(x+t)dt = \arctan t?$$

9. See William Lowell Putnam Mathematical Competition, 1997, Problem A4.

10. A sequence $\{x_n,\ n \geq 1\} \subset \mathbb{R}$ is defined as follows:

$$x_1 = 1, \quad x_{n+1} = \frac{1}{2 + x_n} + \{\sqrt{n}\},\ n \geq 1,$$

V. Brayman and A. Kukush, *Undergraduate Mathematics Competitions (1995–2016)*, Problem Books in Mathematics,
DOI 10.1007/978-3-319-58673-1_4

where $\{a\}$ denotes the fractional part of a. Find the limit

$$\lim_{N\to\infty} \frac{1}{N} \sum_{k=1}^{N} x_k^2.$$

11. See William Lowell Putnam Mathematical Competition, 1995, Problem A5.

12. Let B be a complex Banach space and operators $A, C \in \mathscr{L}(B)$ be such that

$$\sigma(AC^2) \bigcap \{x + iy \mid x + y = 1\} = \varnothing.$$

Prove that

$$\sigma(CAC) \bigcap \{x + iy \mid x + y = 1\} = \varnothing.$$

THE PROBLEMS ARE PROPOSED BY A. A. Dorogovtsev (10), A. Ya. Dorogovtsev (12), A. G. Kukush (8), and V. S. Mazorchuk (4).

1999

Problems 1–9 **for 1–2-Years Students and Problems** 5–11 **for 3–4-Years Students**

1. See Problem 4, 1997.

2. Find the global maximum of a function $2^{\sin x} + 2^{\cos x}$.

3. See William Lowell Putnam Mathematical Competition, 1998, Problem A3.

4. See William Lowell Putnam Mathematical Competition, 1988, Problem A6.

5. See William Lowell Putnam Mathematical Competition, 1998, Problem B5.

6. See William Lowell Putnam Mathematical Competition, 1962, Morning Session, Problem 6.

7. See Problem 5, 1997.

8. See William Lowell Putnam Mathematical Competition, 1961, Morning Session, Problem 7.

9. Let $\{S_n,\ n \geq 1\}$ be a sequence of $m \times m$ matrices such that $S_n S_n^{\mathrm{T}}$ tends to the identity matrix. Prove that there exists a sequence $\{U_n,\ n \geq 1\}$ of orthogonal matrices such that $S_n - U_n \to O$, as $n \to \infty$.

10. Let ξ and η be independent random variables such that $\mathsf{P}(\xi = \eta) > 0$. Prove that there exists a real number a such that $\mathsf{P}(\xi = a) > 0$ and $\mathsf{P}(\eta = a) > 0$.

11. Find a set of linearly independent elements $\mathscr{M} = \{e_i,\ i \geq 1\}$ in an infinite-dimensional separable Hilbert space H, such that the closed linear hull of $\mathscr{M} \setminus \{e_i\}$ coincides with H for every $i \geq 1$

PROBLEM 2 IS PROPOSED BY A. G. Kukush.

V. Brayman and A. Kukush, *Undergraduate Mathematics Competitions (1995–2016)*, Problem Books in Mathematics,
DOI 10.1007/978-3-319-58673-1_5

2000

Problems for 1–2-Years Students

1. Let $\{a_n,\ n \geq 1\}$ be an arbitrary sequence of positive numbers. Denote by b_n the number of terms a_k such that $a_k \geq \frac{1}{n}$. Prove that at least one of the series $\sum_{n=1}^{\infty} a_n$ and $\sum_{n=1}^{\infty} \frac{1}{b_n}$ is divergent.

2. Let $\{M_\alpha,\ \alpha \in \mathscr{A}\}$ be a class of subsets of $\mathbb{N}$ such that for every $\alpha_1, \alpha_2 \in \mathscr{A}$ it holds either $M_{\alpha_1} \subset M_{\alpha_2}$ or $M_{\alpha_2} \subset M_{\alpha_1}$, and moreover $M_{\alpha_1} \neq M_{\alpha_2}$ for each $\alpha_1 \neq \alpha_2$. Is it possible that $\mathscr{A}$ is uncountable?

3. Find all strictly increasing functions $f : [0, +\infty) \to \mathbb{R}$ such that for every $x > y \geq 0$, the equality $f\left(\frac{x+y}{x-y}\right) = \frac{f(x)+f(y)}{f(x)-f(y)}$ holds.

4. A sequence $\{x_n,\ n \geq 1\}$ is defined as follows: $x_1 = a$ and $x_{n+1} = x_n^3 - 3x_n$, $n \geq 1$. Find the set of real numbers a for which the sequence converges.

5. Denote by $d(n)$ the number of positive integer divisors of a positive integer n (including 1 and n). Prove that $\sum_{n=1}^{\infty} \frac{d(n)}{n^2} < 4$.

6. Two wolves and a hare run on the surface of a torus

$$\left\{(x, y, z) \mid (\sqrt{x^2 + y^2} - 2000)^2 + z^2 \leq 2000\right\}$$

at a speed not exceeding 1. Initial distances from each wolf to the hare exceed 2000. The wolves will catch the hare if the distance between at least one of them and the

V. Brayman and A. Kukush, *Undergraduate Mathematics Competitions (1995–2016)*, Problem Books in Mathematics,
DOI 10.1007/978-3-319-58673-1_6

hare becomes smaller then 1. The wolves and the hare see one another at any distance. Can the wolves catch the hare in finite time?

7. In the ring $\mathbb{Z}_n$ of residues modulo n, calculate determinants of matrices A_n and B_n, where $A_n = (\overline{i} + \overline{j})_{\overline{i},\overline{j}=0,1,\ldots,n-1}$, $B_n = (\overline{i} \cdot \overline{j})_{\overline{i},\overline{j}=1,\ldots,n-1}$, $n \geq 2$.

8. Prove that a complex number z satisfies $|z| - \operatorname{Re} z \leq \frac{1}{2}$ if and only if there exist complex numbers u, v such that $z = uv$ and $|u - \overline{v}| \leq 1$.

9. Two (not necessarily distinct) subsets A_1 and A_2 are selected randomly from the class of all subsets of $X = \{1, 2, \ldots, n\}$. Calculate the probability that $A_1 \bigcap A_2 = \varnothing$.

10. There are N chairs in the first row of the Room 41. Assume that all possible ways for n persons to choose their places are equally possible. Calculate the probability that no two persons are sitting alongside.

Problems for 3–4-Years Students

11. Compare the integrals $\int_0^1 x^x dx$ and $\int_0^1 \int_0^1 (xy)^{xy} dxdy$.

12. A sequence $\{x_n,\ n \geq 1\}$ is defined as follows: $x_1 = a$ and $x_{n+1} = 3x_n - x_n^3$, $n \geq 1$. Find the set of real numbers a for which the sequence converges.

13. An element x of a finite group G, $|G| > 1$, is called *self-double* if there exist non-necessarily distinct elements $u \neq e, v \neq e \in G$ such that $x = uv = vu$. Prove that if $x \in G$ is not self-double then x has order 2 and G contains $2(2k-1)$ elements for some $k \in \mathbb{N}$.

14. Find the number of homomorphisms of the rings $M_2(\mathbb{C}) \to M_3(\mathbb{C})$, such that the image of the 2×2 identity matrix is the 3×3 identity matrix.

15. Prove that the system of differential equations

$$\begin{cases} \dfrac{dx}{dt} = y^2 - xy, \\ \dfrac{dy}{dt} = x^4 - x^3 y \end{cases}$$

has no nonconstant periodic solution.

16. A function f satisfies the Lipschitz condition in a neighborhood of the origin in $\mathbb{R}^n$ and $f(\overrightarrow{0}) = \overrightarrow{0}$. Denote by $x(t, t_0, x_0)$, $t \geq t_0$, the solution to Cauchy problem for the system $\frac{dx}{dt} = f(x)$ under initial condition $x(t_0) = x_0$. Prove that:

(a) If zero solution $x(t, t_0, \overrightarrow{0})$, $t \geq t_0$, is stable in the sense of Lyapunov for some $t_0 \in \mathbb{R}$, then it is stable in the sense of Lyapunov for every $t_0 \in \mathbb{R}$ and uniformly in t_0.

(b) If zero solution $x(t, t_0, \overrightarrow{0})$, $t \geq t_0$, is asymptotically stable in the sense of Lyapunov then it holds $\lim\limits_{t\to+\infty} \|x(t, t_0, x_0)\| = 0$ uniformly in x_0 from some neighborhood of the origin in $\mathbb{R}^n$.

17. A function $f\colon [1, +\infty) \to [0, +\infty)$ is Lebesgue measurable, and $\int_1^\infty f(x) \, d\lambda(x) < \infty$ (here λ denotes the Lebesgue measure). Prove that:

(a) the series $\sum_{n=1}^{\infty} f(nx)$ converges for λ-almost all $x \in [1, +\infty)$.

(b) $\lim_{T \to +\infty} \frac{1}{T} \int_1^T x f(x) d\lambda(x) = 0$.

18. Let ξ be a nonnegative random variable. Suppose that for every $x \geq 0$, the expectations $f(x) = \mathsf{E}(\xi - x)_+ \leq \infty$ are known. Evaluate the expectation $\mathsf{E}e^{\xi}$. (Here y_+ denotes $\max(y, 0)$.)

19. The number of passengers at the bus stop is a homogeneous Poisson process with parameter λ, which starts at zero moment. A bus has arrived at time t. Find the expectation of the sum of waiting times for all the passengers.

20. See Problem 10.

THE PROBLEMS ARE PROPOSED BY A.G. Kukush (4, 12, 18), V.S. Mazorchuk (7, 13, 14), Yu.S. Mishura (17), V.M. Radchenko (3, 8, 11), G.M. Shevchenko (1, 5, 6), I.O. Shevchuk (2), O.M. Stanzhytskyi (15, 16), and M.Y. Yadrenko (9, 10, 19, 20).

2001

Problems for 1–2-Years Students

1. Is it true that $\lim_{n\to\infty} |n \sin n| = +\infty$?

2. Let $f \in C^{(2)}(\mathbb{R})$.
(a) Prove that there exists $\theta \in \mathbb{R}$ such that $f(\theta)f''(\theta) + 2(f'(\theta))^2 \geq 0$.
(b) Prove that there exists a function $G : \mathbb{R} \to \mathbb{R}$ such that

$$(\forall x \in \mathbb{R} \;\; f(x)f''(x) + 2(f'(x))^2 \geq 0) \iff G(f(x)) \text{ is convex on } \mathbb{R}.$$

3. Prove that the sequence

$$a_n = \frac{3}{2} \cdot \frac{5}{4} \cdot \frac{9}{8} \cdot \ldots \cdot \frac{2^n + 1}{2^n}$$

converges to some number $a \in (\frac{3}{2}\sqrt[4]{e}, \frac{3}{2}\sqrt{e})$.

4. Find all complex solutions of a system of equations

$$x_1^k + x_2^k + \ldots + x_n^k = 0, \; k = 1, 2, \ldots, n.$$

5. Let A be a nonsingular matrix. Prove that if $\operatorname{rk} A = \operatorname{rk} \left(\begin{smallmatrix} A & B \\ C & D \end{smallmatrix}\right)$ then $D = CA^{-1}B$.

6. Denote by $b(n, k)$ the number of permutations of n elements in which exactly k elements are fixed points. Calculate $\sum_{k=1}^{n} b(n, k)$.

V. Brayman and A. Kukush, *Undergraduate Mathematics Competitions (1995–2016)*, Problem Books in Mathematics,
DOI 10.1007/978-3-319-58673-1_7

Problems for 3–4-Years Students

7. See Problem 4.

8. Let $A(t)$ be $n \times n$ matrix which is continuous in t on $[0, +\infty)$. Let $B \subset \mathbb{R}^n$ be a set of initial values $x(0)$ for which the solution $x(t)$ to a system $\frac{dx}{dt} = A(t)x$ is bounded on $[0, +\infty)$. Prove that B is a subspace of $\mathbb{R}^n$, and if for every $f \in C([0, +\infty), \mathbb{R}^n)$ the system

$$\frac{dx}{dt} = A(t)x + f(t) \tag{$*$}$$

has a bounded on $[0, +\infty)$ solution, then for every $f \in C([0, +\infty), \mathbb{R}^n)$, there exists a unique solution $x(t)$ to $(*)$ which is bounded on $[0, +\infty)$ and satisfies $x(0) \in B^{\perp}$. (Here $B^{\perp}$ denotes the orthogonal complement of B.)

9. Let σ be a random permutation of the set $1, 2, \ldots, n$. (The probability of each permutation is $\frac{1}{n!}$.) Find the expectation of number of the elements which are fixed points of the permutation σ.

10. Find all analytic on $\mathbb{C} \setminus \{0\}$ functions such that the image of every circle with center at zero lies on some circle with center at zero.

11. A cone in $\mathbb{R}^n$ is a set obtained by shift and rotation from the set

$$\{(x_1, \ldots, x_n) : \; x_1^2 + \ldots + x_{n-1}^2 \le r x_n^2\}$$

for some $r > 0$. Prove that if A is an unbounded convex subset of $\mathbb{R}^n$ which does not contain any cone, then there exists a two-dimensional subspace $B \subset \mathbb{R}^n$ such that the projection of A onto B does not contain any cone in $\mathbb{R}^2$.

12. Let $\{\gamma_k, \; k \ge 1\}$ be independent standard Gaussian random variables. Prove that

$$\frac{\max\limits_{1 \le k \le n} \gamma_k^2}{\sum\limits_{k=1}^{n} \gamma_k^2} : \frac{\ln n}{n} \xrightarrow{\mathsf{P}} 2, \;\text{ as } n \to \infty.$$

The problems are proposed by A.G. Kukush (2, 3, 12), A.S. Oliynyk (4, 5, 7), V.M. Radchenko (1), G.M. Shevchenko (6, 9–11), and O.M. Stanzhytskyi (8).

2002

Problems for 1–2-Years Students

1. Does there exist a function $F:\mathbb{R}^2 \to \mathbb{N}$ such that the equality $F(x, y) = F(y, z)$ holds if and only if $x = y = z$?

2. Consider graphs of functions $y = a^{\sin x} + a^{\cos x}$, $x \in \mathbb{R}$, where $a \in [1, 2.5]$. Prove that there exists a point M such that the distance from M to each of the graphs is less than 0.4.

3. Consider a function $f \in C^{(1)}([-1, 1])$, for which $f(-1) = f(1) = 0$. Prove that

$$\exists x \in [-1, 1] : f(x) = (1 + x^2) f'(x).$$

4. Each entry of a matrix $A = (a_{ij}) \in M_n(\mathbb{R})$ is equal to 0 or 1, and moreover $a_{ii} = 0$, $a_{ij} + a_{ji} = 1$, $1 \le i < j \le n$. Prove that $\operatorname{rk} A \ge n - 1$.

5. Prove the inequality

$$\int_0^{\frac{\pi}{2}} \frac{(\cos x)^{\sin x}}{(\cos x)^{\sin x} + (\sin x)^{\cos x}} \, dx < 1.$$

6. Find the dimension of the subspace of linear operators φ on $M_n(\mathbb{R})$ which satisfy $\varphi(A^T) = (\varphi(A))^T$ for every matrix $A \in M_n(\mathbb{R})$.

7. For every $k \in \mathbb{N}$ prove that

$$a_k = \sum_{j=1}^{\infty} \frac{j^k}{j!} \notin \mathbb{Q}.$$

V. Brayman and A. Kukush, *Undergraduate Mathematics Competitions (1995–2016)*, Problem Books in Mathematics,
DOI 10.1007/978-3-319-58673-1_8

8. Find all the functions $f \in C(\mathbb{R})$ such that for every $x, y, z \in \mathbb{R}$ it holds

$$f(x)+f(y)+f(z)=f\left(\tfrac{3}{7}x+\tfrac{6}{7}y-\tfrac{2}{7}z\right)+f\left(\tfrac{6}{7}x-\tfrac{2}{7}y+\tfrac{3}{7}z\right)+f\left(-\tfrac{2}{7}x+\tfrac{3}{7}y+\tfrac{6}{7}z\right).$$

9. Construct a set $A \subset \mathbb{R}$ and a function $f : A \to \mathbb{R}$ such that

$$\forall a_1, a_2 \in A \quad |f(a_1) - f(a_2)| \le |a_1 - a_2|^3,$$

and the range of f is uncountable.

10. Prismatoid is a convex polyhedron such that all its vertices lie in two parallel planes, which are called bases. Given a prismatoid, consider its cross-section which is parallel to the bases and lies at a distance x from the lower base. Prove that the area of this cross-section is a polynomial of x of at most second degree.

Problems for 3–4-Years Students

11. Let ξ be a random variable with finite expectation at a probability space $(\Omega, \mathcal{F}, \mathsf{P})$. Let ω be a signed measure on $\mathcal{F}$ such that

$$\forall A \in \mathcal{F} : \inf_{x \in A} \xi(x) \cdot \mathsf{P}(A) \le \omega(A) \le \sup_{x \in A} \xi(x) \cdot \mathsf{P}(A).$$

Prove that

$$\forall A \in \mathcal{F} : \omega(A) = \int_A \xi(x) d\mathsf{P}(x).$$

12. For every positive integer n consider a function $f_n(x) = n^{\sin x} + n^{\cos x}$, $x \in \mathbb{R}$. Prove that there exists a sequence $\{x_n\}$ such that for every n the function f_n has a global maximum at x_n, and $x_n \to 0$, as $n \to \infty$.

13. Let U be a nonsingular real $n \times n$ matrix, $a \in \mathbb{R}^n$, and L be a subspace of $\mathbb{R}^n$. Prove that

$$\|P_{U^T L}(U^{-1}a)\| \le \|U^{-1}\| \cdot \|P_L a\|,$$

where P_M is the projector onto a subspace M.

14. Let $f:\mathbb{C}\setminus\{0\} \to (0, +\infty)$ be a continuous function, $\lim_{z \to 0} f(z) = 0$, $\lim_{|z| \to \infty} f(z) = \infty$. Prove that for every $T > 0$ there exists a solution to a differential equation $\frac{dz}{dt} = izf(z)$ with a period T.

15. See Problems 5.

16. See Problems 6.

17. See Problems 7.

18. See Problems 8.

19. See Problems 9.

20. See Problems 10.

THE PROBLEMS ARE PROPOSED BY V.B. Brayman (8, 9, 18, 19), V.B. Brayman and Yu.V. Shelyazhenko (7, 17), A.G. Kukush (2, 5, 11–13, 15), A.G. Kukush and R.P. Ushakov (10, 20), A.S. Oliynyk (4, 6, 16), A.V. Prymak (3), and O.M. Stanzhytskyi (14).

2003

Problems 1–8 for 1–2-Years Students and problems 5–12 for 3–4-Years Students

1. Evaluate

$$\sum_{n=1}^{\infty} \frac{9n+4}{n(3n+1)(3n+2)}.$$

2. Find the limit

$$\lim_{N\to\infty} \sqrt{N}\left(1 - \max_{1\le n\le N} \{\sqrt{n}\}\right),$$

where $\{x\}$ denotes the fractional part of x.

3. For every $n \in \mathbb{N}$, find the minimal $k \in \mathbb{N}$ for which there exist $\overrightarrow{x_1}, \ldots, \overrightarrow{x_k} \in \mathbb{R}^n$ such that

$$\forall\, \vec{x} \in \mathbb{R}^n\ \exists\, a_1, \ldots, a_k > 0 : \vec{x} = \sum_{i=1}^{k} a_i \overrightarrow{x_i}.$$

4. For which $n \in \mathbb{N}$ there exist $n \times n$ matrices A and B such that $\operatorname{rk} A + \operatorname{rk} B \le n$ and every square real matrix X which commutes with A and B is proportional to the identity matrix (i.e., it has a form $X = \lambda I$, $\lambda \in \mathbb{R}$)?

5. Prove the inequality

$$\sqrt{2\sqrt[3]{3\sqrt[4]{4\ldots\sqrt[n]{n}}}} < 2,\ n \ge 2.$$

V. Brayman and A. Kukush, *Undergraduate Mathematics Competitions (1995–2016)*, Problem Books in Mathematics,
DOI 10.1007/978-3-319-58673-1_9

6. For every real $x \neq 1$ find the sum

$$\sum_{n=0}^{\infty} \frac{x^{3^n} + \left(x^{3^n}\right)^2}{1 - x^{3^{n+1}}}.$$

7. For every positive integers $m \leq n$ prove the inequality

$$\sum_{k=0}^{m} (-1)^{m+k} \binom{m}{k} \left(\frac{k}{m}\right)^n \leq \binom{n}{m} \frac{m!}{m^m}.$$

8. A parabola with focus F and a triangle T are drawn in the plane. Using a compass and a ruler, construct a triangle similar to T such that one of its vertices is F and other two vertices lie on the parabola.

9. Does there exist a Lebesgue measurable set $A \subset \mathbb{R}^2$ such that for every set E of zero Lebesgue measure the set $A \backslash E$ is not Borel measurable?

10. A real symmetric matrix $A = (a_{ij})_{i,j=1}^n$ with eigenvectors $\{e_k,\ 1 \leq k \leq n\}$ and eigenvalues λ_k, $1 \leq k \leq n$, is given. Construct a real symmetric positive semidefinite matrix $X = (x_{ij})_{i,j=1}^n$ which minimizes the distance $d(X, A) = \sqrt{\sum_{i,j=1}^{n} (x_{ij} - a_{ij})^2}$.

11. Let φ be a conform mapping from $\Omega = \{\operatorname{Im} z > 0\} \backslash T$ onto $\{\operatorname{Im} z > 0\}$, where T is a triangle with vertices $\{1, -1, i\}$. Point $z_0 \in \Omega$ is such that $\varphi(z_0) = z_0$. Prove that $|\varphi'(z_0)| \geq 1$.

12. The vertices of a triangle are independent random points uniformly distributed at a unit circle. Find the expectation of the area of this triangle.

THE PROBLEMS ARE PROPOSED BY T. O. Androshchuk (11), A. V. Bondarenko (3, 4, 9), A. G. Kukush (1, 5, 6, 10, 12), D. Yu. Mitin (2, 7), and G. M. Shevchenko (8).

2004

Problems for 1–2-Years Students

1. Prove that for every positive integer n the inequality

$$\frac{1}{3!}+\frac{3}{4!}+\ldots+\frac{2n-1}{(n+2)!}<\frac{1}{2}$$

holds.

2. One cell is erased from the $2\times n$ table in arbitrary way. Find the probability of the following event: It is possible to cover the rest of the table with figures ⊞ of any orientation without overlapping.

3. For every continuous and convex on $[0,1]$ function f prove the inequality

$$\frac{2}{5}\int_0^1 f(x)\,dx+\frac{2}{3}\int_0^{3/5} f(x)\,dx\ge\int_0^{4/5} f(x)\,dx.$$

4. Find all odd continuous functions $f:\mathbb{R}\to\mathbb{R}$ such that the equality $f(f(x))=x$ holds for every real x.

5. Using a compass and a ruler, construct a circle of the maximal radius which lies inside the given parabola and touches it in its vertex.

6. Let A, B, C, and D be (not necessarily square) real matrices such that

$$A^{\mathrm{T}}=BCD,\ B^{\mathrm{T}}=CDA,\ C^{\mathrm{T}}=DAB,\ D^{\mathrm{T}}=ABC.$$

For the matrix $S=ABCD$ prove that $S^3=S$.

V. Brayman and A. Kukush, *Undergraduate Mathematics Competitions (1995–2016)*, Problem Books in Mathematics,
DOI 10.1007/978-3-319-58673-1_10

7. Denote by A_n the maximal determinant of $n \times n$ matrix with entries ± 1. Does there exist a finite limit $\lim\limits_{n\to\infty} \sqrt[n]{A_n}$?

8. Let $\{x_n,\ n \geq 1\}$ be a sequence of positive numbers which contains at least two distinct elements. Is it always

$$\liminf_{n\to\infty} \left(x_1 + \ldots + x_n - n\sqrt[n]{x_1 \ldots x_n}\right) > 0?$$

9. A permutation of the entries of matrix maps each nonsingular $n \times n$ matrix into a nonsingular one and maps the identity matrix into itself. Prove that the permutation preserves the determinant of a matrix.

10. A rectangle with side lengths a_0 and b_0 is dissected into smaller rectangles with side lengths a_k and b_k, $1 \leq k \leq n$. The sides of the smaller rectangles are parallel to the corresponding sides of the big rectangle. Prove that

$$|\sin a_0 \sin b_0| \leq \sum_{k=1}^{n} |\sin a_k \sin b_k|.$$

Problems for 3–4-Years Students

11. A random variable ξ is distributed as $|\gamma|^\alpha$, $\alpha \in \mathbb{R}$, where γ is a standard normal variable. For which α does there exist $\mathsf{E}\xi$?

12. See Problem 2.

13. A normed space Y is called strictly normed if for every $y_1, y_2 \in Y$ the equality $\|y_1\| = \|y_2\| = \|\frac{y_1+y_2}{2}\|$ implies $y_1 = y_2$. Let X be a normed space, G be a subspace of X and the adjoint space X^* be strictly normed. Prove that for every functional from G^* there exists a unique extension in X^* which preserves the norm.

14. Let $R(z) = \dfrac{z^2}{2} - z + \ln(1+z)$, $z \in \mathbb{C}$, $z \neq -1$. Prove that for every real x the inequality $|R(ix)| \leq \dfrac{|x|^3}{3}$ holds.
(Here "ln" means the value of the logarithm from the branch with $\ln 1 = 0$.)

15. Let A, B, C, and D be (not necessarily square) real matrices such that

$$A^{\mathrm{T}} = BCD,\ B^{\mathrm{T}} = CDA,\ C^{\mathrm{T}} = DAB,\ D^{\mathrm{T}} = ABC.$$

For $S = ABCD$ prove that $S^2 = S$.
Remark: for 1–2-years students it is proposed to prove that $S^3 = S$.

16. Let e be a nonzero vector in $\mathbb{R}^2$. Construct a nonsingular matrix $A \in \mathbb{R}^{2\times 2}$ such that for $f_d(x) := \|A(x+d)\|^2$, $x, d \in \mathbb{R}^2$, there exist at least 8 couples of points

(x, y) such that $f_e(x) = 1$, $f_{-e}(y) = 1$, and moreover there exist real numbers λ and μ such that (x, y) is a stationary point of Lagrange function

$$F(x, y) := \|x - y\|^2 + \lambda f_e(x) + \mu f_{-e}(y).$$

17. See Problem 9.

18. A croupier and two players play the following game. The croupier chooses an integer in the interval $[1, 2004]$ with uniform probability. The players guess the integer in turn. After each guess, the croupier informs them whether the chosen integer is higher or lower or has just been guessed. The player who guesses the integer first wins. Prove that both players have strategies such that their chances to win are at least $\frac{1}{2}$.

19. See Problem 10.

20. Does there exist a sequence $\{x_n, n \geq 1\}$ of vectors from l_2 with unit norm satisfying $(x_n, x_m) < -\dfrac{1}{2004}$ for $n \neq m$, $n, m \in \mathbb{N}$?

The problems are proposed by A.V. Bondarenko (7, 20), V.B. Brayman (6, 9, 15, 17), Zh.T. Chernousova (5), A.G. Kukush (2, 4, 12, 16), Yu.S. Mishura (11), D.Yu. Mitin (8, 14), O.N. Nesterenko (13), Zsolt Páles (Hungary) (10, 19), A.V. Prymak (3), S.V. Shklyar (18), and R.P. Ushakov (1).

2005

Problems for 1–2-Years Students

1. Is it true that a sequence $\{x_n, n \geq 1\}$ of real numbers converges if and only if

$$\lim_{n\to\infty} \limsup_{m\to\infty} |x_n - x_m| = 0?$$

2. Let A, B, and C be real matrices of the same size. Prove the inequality

$$\operatorname{tr}(A(A^T - B^T) + B(B^T - C^T) + C(C^T - A^T)) \geq 0.$$

3. A billiard table is obtained by cutting out some squares from the chessboard. The billiard ball is shot from one of the table corners in such a way that its trajectory forms angle α with the side of the billiard table, $\tan\alpha \in \mathbb{Q}$. When the ball hits the border of the billiard table it reflects according to the rule: the incidence angle equals the reflection angle. If the ball lands on any corner it falls into a hole. Prove that the ball will necessarily fall into some hole.

4. Solve an equation

$$\lim_{n\to\infty} \sqrt{1 + \sqrt{x + \sqrt{x^2 + \ldots + \sqrt{x^n}}}} = 2.$$

5. Do there exist matrices A, B, and C which have no common eigenvectors and satisfy the condition $AB = BC = CA$?

6. Prove that

$$\int_{-\pi}^{\pi} \cos 2x \cos 3x \cos 4x \ldots \cos 2005x \, dx > 0.$$

V. Brayman and A. Kukush, *Undergraduate Mathematics Competitions (1995–2016)*, Problem Books in Mathematics,
DOI 10.1007/978-3-319-58673-1_11

7. Let $f \in C^{(1)}(\mathbb{R})$ and $a_1 < a_2 < a_3 < b_1 < b_2 < b_3$. Do there always exist real numbers $c_1 \le c_2 \le c_3$ such that $c_i \in [a_i, b_i]$ and

$$f'(c_i) = \frac{f(b_i) - f(a_i)}{b_i - a_i}, \ i = 1, 2, 3?$$

8. Call $\mathbb{Z}$-ball a set of points of the form

$$S = \{(x, y, z) \mid x^2 + y^2 + z^2 \le R^2, \ x, y, z \in \mathbb{Z}\}, \ R \in \mathbb{R}.$$

Prove that there is no $\mathbb{Z}$-ball which contains exactly 2005 distinct points.

9. Consider a triangle $A_1A_2A_3$ at Cartesian plane with sides and their extensions not passing through the origin O. Call such triangle positive if for at least two of numbers $i = 1, 2, 3$ vector $\overrightarrow{OA}$ turns counterclockwise when point A moves from A_i to A_{i+1} (here $A_4 = A_1$), and negative otherwise. Let (x_i, y_i) be coordinates of points A_i, $i = 1, 2, 3$. Prove that there is no polynomial $P(x_1, y_1, x_2, y_2, x_3, y_3)$ which is positive for positive triangles $A_1A_2A_3$ and negative for negative ones.

Problems for 3–4-Years Students

10. Let K be a compact set in the space $C([0, 1])$ with uniform metric. Prove that the function $f(t) = \min\{x(t) + x(1-t) : x \in K\}$, $t \in [0, 1]$ is continuous.

11. Find all $\lambda \in \mathbb{C}$ such that every sequence $\{a_n, n \ge 1\} \subset \mathbb{C}$, which satisfies $|\lambda a_{n+1} - \lambda^2 a_n| < 1$ for each $n \ge 1$, is bounded.

12. Let X and Y be linear normed spaces. An operator $K : X \to Y$ is called supercompact if for every bounded set $M \subset X$ the set

$$K(M) = \{y \in Y \mid \exists x \in M : y = K(x)\}$$

is compact in Y. Prove that among linear continuous operators from X to Y, only zero operator is supercompact.

13. Let A be a real orthogonal matrix such that $A^2 = I$, where I is the identity matrix. Prove A can be written as $A = UBU^T$, where U is an orthogonal matrix and B is a diagonal matrix with entries ± 1 on the diagonal.

14. Let B be a bounded subset of a connected metric space X. Does there always exist a connected and bounded subset $A \subset X$ such that $B \subset A$?

15. Let $t > 0$ and μ be a measure on Borel sigma-algebra of $\mathbb{R}^+$ such that $\int_{\mathbb{R}^+} \exp(\alpha x^t)\, d\mu(x) < \infty$ for every $\alpha < 1$. Prove that $\int_{\mathbb{R}^+} \exp(\alpha(x+1)^t)\, d\mu(x) < \infty$ for every $\alpha < 1$.

16. See Problem 8.

17. See Problem 9.

18. Let $x_0 < x_1 < \ldots < x_n$ and $y_0 < y_1 < \ldots < y_n$. Prove that there exists a strictly increasing on $[x_0, x_n]$ polynomial p such that $p(x_j) = y_j, j = 0, \ldots, n$.

THE PROBLEMS ARE PROPOSED BY A. V. Bondarenko (8, 16), V. B. Brayman (5, 7), V. S. Grinberg (USA) (9, 17), G. V. Kryukova (3), A. G. Kukush (4, 10, 13, 15), O. N. Nesterenko (1), A. V. Prymak (11, 18), M. S. Pupashenko (6, 14), I. O. Senko (12), and M. S. Viazovska (2).

2006

Problems for 1–2-Years Students

1. Find all positive integers n such that the polynomial $(x^4 - 1)^n + (x^2 - x)^n$ is divisible by $x^5 - 1$.

2. Let $z \in \mathbb{C}$ be such that points z^3, $2z^3 + z^2$, $3z^3 + 3z^2 + z$, and $4z^3 + 6z^2 + 4z + 1$ are the vertices of some inscribed quadrangle on the complex plane. Find $\operatorname{Re} z$.

3. Find the minimum of the expression $\max\limits_{1 \le i < j \le n+1} (x_i, x_j)$ over all the unit vectors $x_1, \ldots, x_{n+1} \in \mathbb{R}^n$.

4. Let I_m be the identity $m \times m$ matrix, $A \in M_{m\times n}(\mathbb{R})$, and B be a symmetric $n \times n$ matrix such that the block matrix $\begin{pmatrix} I_m & A \\ A^T & B \end{pmatrix}$ is positive definite. Prove that the "matrix determinant" $B - A^T A$ is positive definite as well.

5. Does there exist an infinite set $\mathcal{M}$ of symmetric matrices such that for every distinct matrices $A, B \in \mathcal{M}$ it holds $AB^2 = B^2 A$ but $AB \neq BA$?

6. Let $f : (0, +\infty) \to \mathbb{R}$ be a continuous concave function, for which $\lim\limits_{x\to+\infty} f(x) = +\infty$ and $\lim\limits_{x\to+\infty} \frac{f(x)}{x} = 0$. Prove that $\sup\limits_{n\in\mathbb{N}} \{f(n)\} = 1$, where $\{a\}$ is the fractional part of a.

7. Does there exist a continuous function $f : \mathbb{R} \to (0, 1)$ such that the sequence $a_n = \int_{-n}^{n} f(x)\,dx$, $n \ge 1$, converges, and the sequence $b_n = \int_{-n}^{n} f(x) \ln f(x)\,dx$, $n \ge 1$, diverges?

V. Brayman and A. Kukush, *Undergraduate Mathematics Competitions (1995–2016)*, Problem Books in Mathematics,
DOI 10.1007/978-3-319-58673-1_12

8. Let $P(x)$ be a polynomial such that there exist infinitely many couples of integers (a, b) for which $P(a + 3b) + P(5a + 7b) = 0$. Prove that the polynomial has an integer root.

9. For every real numbers $a_1, a_2, \ldots, a_n \in \mathbb{R} \setminus \{0\}$ prove the inequality

$$\sum_{i,j=1}^{n} \frac{a_i a_j}{a_i^2 + a_j^2} \geq 0.$$

Problems for 3–4-Years Students

10. See Problem 2.

11. Does there exist a continuous function $f : \mathbb{R} \to (0, 1)$ such that $\int_{-\infty}^{\infty} f(x)\,dx < \infty$ while $\int_{-\infty}^{\infty} f(x) \ln f(x)\,dx$ diverges?

12. See Problem 4.

13. See Problem 5.

14. (a) Random variables ξ and η (not necessarily independent) have a continuous cumulative distribution functions. Prove that $\min(\xi, \eta)$ has a continuous cumulative distribution function as well.

(b) Random variables ξ and η have a probability density functions. Is it true that $\min(\xi, \eta)$ has a probability density function as well?

15. Is it possible to choose an uncountable set $\mathscr{A} \subset l_2$ of elements with unit norm such that for every distinct $x = (x_1, \ldots, x_n, \ldots)$, $y = (y_1, \ldots, y_n, \ldots)$ from $\mathscr{A}$ the series $\sum_{n=1}^{\infty} |x_n - y_n|$ diverges?

16. Let ξ and η be independent identically distributed random variables such that $\mathsf{P}(\xi \neq 0) = 1$. Prove the inequality

$$\mathsf{E}\frac{\xi\eta}{\xi^2 + \eta^2} \geq 0.$$

17. For every $n \in \mathbb{N}$ find the minimal $\lambda > 0$ such that for every convex compact set $K \subset \mathbb{R}^n$ there exists a point $x \in K$ with the following property: a set which is homothetic to K with the center x and coefficient $(-\lambda)$ contains K.

18. Let $X = L_1[0, 1]$ and $T_n : X \to X$ be a sequence of nonnegative (i.e., $f \geq 0 \implies T_n f \geq 0$) continuous linear operators such that $\|T_n\| \leq 1$ and $\lim_{n\to\infty} \|f - T_n f\|_X = 0$, for $f(x) \equiv x$ and for $f(x) \equiv 1$. Prove that $\lim_{n\to\infty} \|f - T_n f\|_X = 0$ for every $f \in X$.

The problems are proposed by A. V. Bondarenko (15), A. V. Bondarenko and M. S. Viazovska (3), V. B. Brayman (2, 5, 8, 10, 13), A. G. Kukush (1, 4, 7, 11, 12), A. G. Kukush and G. M. Shevchenko (14), O. Lytvak (Canada) (17), O. N. Nesterenko (6), S. Novak (Great Britain) (9, 16), and A. V. Prymak (18).

2007

Problems for 1–2-Years Students

1. Let p, q, r, and s be positive integers. Find the limit

$$\lim_{n\to\infty}\prod_{k=1}^{n}\frac{(k+p)(k+q)}{(k+r)(k+s)}.$$

2. Is it true that for every $n \ge 2$ the number $\sum_{k=1}^{n} k\binom{2n}{k}$ is divisible by 8?

3. Two players in turn replace asterisks in the matrix $\begin{pmatrix} * & * & \dots & * \\ * & * & \dots & * \\ \dots & \dots & \dots & \dots \\ * & * & \dots & * \end{pmatrix}$ of size 10×10 with positive integers $1, \dots, 100$ (at each step, one can put any number which had not been used earlier instead of any asterisk). If they form a nonsingular matrix then the first player wins, otherwise the second player wins. Has any of the players a winning strategy? If somebody has, then who?

4. Prove that a function $f \in C^{(1)}((0, +\infty))$, for which

$$f'(x) = \frac{1}{1 + x^4 + \cos f(x)},\ x > 0,$$

is bounded on $(0, +\infty)$.

5. Does there exist a polynomial which attains the value k exactly at k distinct real points for every $1 \le k \le 2007$?

V. Brayman and A. Kukush, *Undergraduate Mathematics Competitions (1995–2016)*, Problem Books in Mathematics,
DOI 10.1007/978-3-319-58673-1_13

6. The clock-face is a disk of radius 1. The hour-hand is a disk of radius 1/2 is internally tangent to the circle of the clock-face, and the minute-hand is a line segment of length 1. Find the area of the figure formed by all intersections of the hands in 12 hours (i.e., in one full turn of the hour-hand).

7. Find the maximum of $x_1^3 + \ldots + x_{10}^3$ for $x_1, \ldots, x_{10} \in [-1, 2]$ such that

$$x_1 + \ldots + x_{10} = 10.$$

8. Let $a_0 = 1$, $a_1 = 1$, and $a_n = a_{n-1} + (n-1)a_{n-2}$, $n \geq 2$. Prove that for every odd number p the number $a_p - 1$ is divisible by p.

9. Find all positive integers n for which there exist infinitely many $n \times n$ matrices A with integer entries such that $A^n = I$ (here I is the identity matrix).

Problems for 3–4-Years Students

10. Does the Riemann integral

$$\int_0^\infty \frac{\sin x \, dx}{x + \ln x}$$

converge?

11. See Problem 6.

12. See Problem 4.

13. See Problem 5.

14. Let $f : \mathbb{R} \to [0, +\infty)$ be a Lebesgue measurable function such that $\int_A f \, d\lambda < +\infty$ for every set A of finite Lebesgue measure. Prove that there exist a constant M and a Lebesgue integrable function $g : \mathbb{R} \to [0, +\infty)$ such that $f(x) \leq g(x) + M$, $x \in \mathbb{R}$.

15. Investigate the character of monotonicity of the function $f(\sigma) = \mathsf{E}\dfrac{1}{1 + e^{\xi}}$, $\sigma > 0$, where ξ is normal random variable with mean m and variance σ^2 (here m is a real parameter).

16. See Problem 7.

17. Let A and B be symmetric real positive definite matrices and the matrix $A + B - I$ be positive definite as well, where I is the identity matrix. Is it possible that the matrix

$$A^{-1} + B^{-1} - \frac{1}{2}(A^{-1}B^{-1} + B^{-1}A^{-1})$$

is negative definite?

18. Let $P(z)$ be a polynomial with leading coefficient 1. Prove that there exists a point z_0 on the unit circle $\{z \in \mathbb{C} : |z| = 1\}$ such that $|P(z_0)| \geq 1$.

THE PROBLEMS ARE PROPOSED BY A. V. Bondarenko and M. S. Viazovska (9), V. B. Brayman (3, 5, 13), A. G. Kukush (2, 10, 15, 17), D. Yu. Mitin (7, 16), O. N. Nesterenko (4, 12), V. M. Radchenko (14), O. V. Rybak (8, 18), G. M. Shevchenko (6, 11), and R. P. Ushakov (1).

2008

Problems for 1–2-Years Students

1. Find $\inf\{a+b+c\ :\ a,b,c>0,\ \sin a\cdot\sin b\cdot\sin c=\frac{3}{\pi}\,abc\}$.

2. Let f be a continuous and bounded on $\mathbb{R}$ function such that

$$\sup_{x\in\mathbb{R}}|f(x+h)-2f(x)+f(x-h)|\to 0,\ \text{as}\ h\to 0.$$

Does it follow that f is uniformly continuous on $\mathbb{R}$?

3. Let f be 4 times differentiable even function on $\mathbb{R}$, and $g(x)=f(\sqrt{x})$, $x\geq 0$. Prove that g is twice differentiable at $x=0$, and find $g''(0)$ in terms of derivatives of f at zero.

4. A function $f:[4,+\infty)\to\mathbb{R}$ satisfies the following conditions:
(a) $f(x^2)=f(x)+[\log_2\log_2 x]^{-2}$, where $[t]$ is the integer part of t;
(b) there exists $\lim\limits_{x\to+\infty}f(x)$.
Prove that f is monotone.

5. A polynomial $P(x)=x^n+p_{n-1}x^{n-1}+\ldots+p_0$ has exactly m $(2\leq m\leq n)$ distinct complex roots. Prove that at least one of the coefficients $p_{n-1},\ \ldots,\ p_{n-m}$ is nonzero.

6. Let A be a complex matrix of size $n\times k$ such that $A^TA=0$. Find the maximal possible rank of A.

7. Let $f\in C^{(\infty)}(\mathbb{R}\setminus\{0\})\cap C(\mathbb{R})$ be such that $f(x)=o(x^n)$, as $x\to 0$, for all $n\in\mathbb{N}$. Is it necessary that $f\in C^{(1)}(\mathbb{R})$?

V. Brayman and A. Kukush, *Undergraduate Mathematics Competitions (1995–2016)*, Problem Books in Mathematics,
DOI 10.1007/978-3-319-58673-1_14

8. For a real square matrix $A = (a_{ij})_{i,j=1}^{n}$ set $A^S = (\widetilde{a}_{ij})$, where

$$\widetilde{a}_{ij} = \begin{cases} a_{ij} & \text{if } i+j \text{ is even,} \\ a_{ji} & \text{if } i+j \text{ is odd.} \end{cases}$$

Find all square matrices A such that for every matrix B of the same size the equality $(AB)^S = B^S A^S$ holds.

9. Some n points with positive integer coordinates are marked at the coordinate plane. It is known that if a point (x, y) is marked then all the points with positive integer coordinates (x', y') such that $x' \leq x$, $y' \leq y$ are marked as well. For every marked point (x, y), denote by $R(x, y)$ the number of marked points (x', y') such that $x' \geq x$ and $y' \geq y$. Prove that there exist at least $n/4$ points (x, y) for which $R(x, y)$ is odd.

Problems for 3–4-Years Students

10. See Problem 5.

11. Let ξ be a random variable such that ξ and ξ^2 are independent. Prove that there exists a real number c such that $\cos \xi = c$, almost surely.

12. See Problem 6.

13. Let f be $2k$ times differentiable even function on $\mathbb{R}$, and $g(x) = f(\sqrt{x})$, $x \geq 0$. Prove that g is k times differentiable at $x = 0$, and find $g^{(k)}(0)$ in terms of derivatives of f at zero.

14. Let $A = (a_{ij})_{i,j=1}^{n}$, $B = (b_{ij})_{i,j=1}^{n}$ be real symmetric matrices and $\lambda_{\min}(A)$, $\lambda_{\min}(B)$ be their smallest eigenvalues. Prove the inequality

$$|\lambda_{\min}(A) - \lambda_{\min}(B)| \leq n \cdot \max_{1 \leq i,j \leq n} |a_{ij} - b_{ij}|.$$

15. Let $f : \mathbb{R}^2 \to \mathbb{R}^2$ be a continuously differentiable mapping such that for every $x \in \mathbb{R}^2$ the matrix $Df(x) + (Df(x))^T$ is nonsingular and $Df(0) = I$ (here $Df(x)$ is the Jacobian matrix at point x and I is the identity matrix). Is it necessary that f is an injection?

16. Let ξ be a random variable with positive probability density function. Is it always true that there exist two distinct functions $f, g \in C(\mathbb{R})$ such that $f(\xi)$ and $g(\xi)$ are identically distributed?

17. Let $f : [0, 1] \to \mathbb{R}$ be a Lebesgue measurable function, λ be Lebesgue measure on $[0, 1]$. It is known that for every open set $A \subset [0, 1]$ it holds

$$\int_A f^{2n-1}(x)\, d\lambda(x) \to 0, \text{ as } n \to \infty.$$

Prove that $\lambda(\{x : |f(x)| \geq 1\}) = 0$.

18. Let M be the set of all nonsingular 3×3 matrices over the field $\mathbb{Z}_2$. Find the smallest positive integer n such that $A^n = I$ for all $A \in M$.

The problems are proposed by A.V. Bondarenko (18), A.V. Bondarenko and A.V. Prymak (2), V.B. Brayman (1, 8), V.B. Brayman and O.V. Rudenko (6, 12), O.V. Ivanov and A.G. Kukush (14), A.G. Kukush (3, 4, 15), A.G. Kukush and V.B. Brayman (13), O.N. Nesterenko (7), S. Novak (Great Britain) (11, 16), V.M. Radchenko (17), and O.V. Rybak (5, 9, 10).

2009

Problems for 1–2-Years Students

1. Triangle ABC is inscribed into a circle. Does there always exist a point D on this circle such that $ABCD$ is a circumscribed quadrilateral?

2. Let $F_0 = 0,\ F_1 = 1,\ F_k = F_{k-1} + F_{k-2},\ k \geq 2$ be the Fibonacci numbers. Find all positive integers n for which the polynomial $F_n x^{n+1} + F_{n+1} x^n - 1$ is irreducible in the ring $\mathbb{Q}[x]$ of polynomials with rational coefficients.

3. Let A, B, and C be the angles of an acute triangle. Prove the inequalities

$$\text{(a)}\quad \frac{\cos A}{\sin B \sin C} + \frac{\cos B}{\sin C \sin A} + \frac{\cos C}{\sin A \sin B} \geq 2;$$

$$\text{(b)}\quad \frac{\cos A}{\sqrt{\sin B \sin C}} + \frac{\cos B}{\sqrt{\sin C \sin A}} + \frac{\cos C}{\sqrt{\sin A \sin B}} \leq \sqrt{3}.$$

4. Find all the positive integers n for which there exist matrices $A, B, C \in M_n(\mathbb{Z})$ such that $ABC + BCA + CAB = I$. Here I is the identity matrix.

5. Let $x, y : \mathbb{R} \to \mathbb{R}$ be a couple of functions such that

$$\forall t, s \in \mathbb{R} \quad (x(t) - x(s))\,(y(t) - y(s)) \geq 0.$$

Prove that there exist two nondecreasing functions $f, g : \mathbb{R} \to \mathbb{R}$ and a function $z : \mathbb{R} \to \mathbb{R}$ such that $x(t) = f\,(z(t))$ and $y(t) = g\,(z(t))$ for all $t \in \mathbb{R}$.

6. Let $\{x_n,\ n \geq 1\}$ be a sequence of real numbers such that there exists a finite limit $\lim\limits_{n\to\infty} \frac{1}{n} \sum\limits_{k=1}^{n} x_k$. Prove that for every $p > 1$ there exists a finite limit $\lim\limits_{n\to\infty} \frac{1}{n^p} \sum\limits_{k=1}^{n} k^{p-1} x_k$.

V. Brayman and A. Kukush, *Undergraduate Mathematics Competitions (1995–2016)*, Problem Books in Mathematics,
DOI 10.1007/978-3-319-58673-1_15

7. Let $K(x) = xe^{-x}$, $x \in \mathbb{R}$. For every $n \geq 3$ find

$$\sup_{x_1,\ldots,x_n \in \mathbb{R}} \min_{1 \leq i < j \leq n} K(|x_i - x_j|).$$

8. Does there exist a function $f : \mathbb{Q} \to \mathbb{Q}$ such that $f(x)f(y) \leq |x - y|$ for every $x, y \in \mathbb{Q}$, $x \neq y$, and for each $x \in \mathbb{Q}$ the set $\{y \in \mathbb{Q} \mid f(x)f(y) = |x - y|\}$ is infinite?

9. Find all $n \geq 2$ for which it is possible to enumerate all permutations of the set $\{1, \ldots, n\}$ with numbers $1, \ldots, n!$ in such a way that for each couple of permutations σ, τ with adjacent indices, as well as for the couple of permutations with indices 1 and $n!$, it holds $\sigma(k) \neq \tau(k)$ for every $1 \leq k \leq n$.

Problems for 3–4-Years Students

10. See Problem 4.

11. See Problem 6.

12. See Problem 5.

13. See Problem 7.

14. See Problem 8.

15. Let μ be a measure on Borel sigma-algebra in $\mathbb{R}$ such that

$$\forall a \in \mathbb{R} \quad \int_{\mathbb{R}} e^{ax} d\mu(x) < \infty,$$

moreover $\mu((-\infty, 0)) > 0$ and $\mu((0, +\infty)) > 0$. Prove that there exists a unique real a such that $\int_{\mathbb{R}} xe^{ax} d\mu(x) = 0$.

16. Let $\{\xi_n, n \geq 0\}$ and $\{\nu_n, n \geq 1\}$ be two independent sets of independent identically distributed random variables (here the probability distributions could be different in the first and the second sets). It is known that $\mathsf{E}\xi_0 = 0$ and $\mathsf{P}\{\nu_1 = 1\} = p$, $\mathsf{P}\{\nu_1 = 0\} = 1 - p$, $p \in (0, 1)$. Let $x_0 = 0$ and $x_n = \sum_{k=1}^{n} \nu_k$, $n \geq 1$. Prove that $\frac{1}{n} \sum_{k=0}^{n} \xi_{x_k} \to 0$, almost surely, as $n \to \infty$.

17. Let $X_1, \ldots, X_{2n}$ be a set of independent identically distributed random variables such that $X_1 \neq 0$ almost surely. Define

$$Y_k = \frac{\left|\sum_{i=1}^{k} X_i\right|}{\sqrt{\sum_{i=1}^{k} X_i^2}}, \quad 1 \leq k \leq 2n.$$

Prove the inequality $\mathsf{E}(Y_{2n}^2) \leq 1 + 4(\mathsf{E}Y_n)^2$.

18. See Problem 9.

THE PROBLEMS ARE PROPOSED BY A. V. Bondarenko and E. Saff (USA) (7, 13), V. B. Brayman (4, 8, 10, 14), J. Dhaene (Belgium) and V. B. Brayman (5, 12), A. A. Dorogovtsev (16), A. G. Kukush (1, 15), A. G. Kukush and M. M. Rozhkova (3), D. Yu. Mitin (6, 11), S. Novak (Great Britain) (17), O. B. Rudenko (9, 18), and R. P. Ushakov (2).

2010

Problems for 1–2-Years Students

1. Does there exist a family of functions $f_\alpha : [0, 1] \to \mathbb{R}$, $\alpha \in \mathbb{R}$, such that the intersection of graphs of any two distinct functions from the family contains exactly 3 points, while the intersection of graphs of any three distinct functions contains exactly 2 points?

2. A sequence $\{a_n,\ n \ge 1\}$ satisfies the condition $[(n+1)a_n] = [na_{n+1}]$, $n \ge 1$ (here $[x]$ is the integer part of x). Prove that there exists $c \in \mathbb{R}$ such that the inequality $|a_n - cn| < 1$ holds for every $n \ge 1$.

3. Functions $f, g \in C([a, b])$ are differentiable on (a, b), and $g(x) \neq 0$, $x \in [a, b]$. Prove that there exists $x \in (a, b)$ such that

$$f'(x)(ag(b) - bg(a)) - g'(x)(af(b) - bf(a)) = \left(\frac{f}{g}\right)'(x)(a-b)g(a)g(b).$$

4. Let $S = S(A_1, \dots, A_k)$ be the smallest set of real matrices of size $n \times n$, $n \ge 2$, which has the following properties:

(a) $A_1, \dots, A_k \in S$;

(b) if $A, B \in S$ and $\alpha, \beta \in \mathbb{R}$ then $\alpha A + \beta B \in S$ and $AB \in S$.

For which minimal k do there exist matrices $A_1, \dots, A_k$ such that $S(A_1, \dots, A_k) = M_n(\mathbb{R})$?

5. A sequence $\{x_n,\ n \ge 1\}$ satisfies $x_{n+1} = x_n + e^{-x_n}$, $n \ge 1$, and $x_1 = 1$. Find $\lim\limits_{n\to\infty} \frac{x_n}{\ln n}$.

6. Let n be a fixed positive integer. Find the smallest k such that for any real numbers a_{ij}, $1 \le i, j \le n$, there exists a real polynomial $P(x, y)$ of degree at most k such that $P(i, j) = a_{ij}$ for every $1 \le i, j \le n$.

V. Brayman and A. Kukush, *Undergraduate Mathematics Competitions (1995–2016)*, Problem Books in Mathematics, DOI 10.1007/978-3-319-58673-1_16

7. Let A be a symmetric singular matrix of size $n \times n$, $n \geq 2$, with integer entries. Denote by A_i the matrix obtained from A by erasing ith row and ith column, $1 \leq i \leq n$. Assume that $\det A_1 = 2010$. Prove that $\det A_i$ is divisible by 2010 for every $2 \leq i \leq n$.

8. Points $P_1, \ldots, P_n$ are chosen at a plane such that not all of them are collinear. For every $1 \leq i, j, k \leq n$, $i \neq j$, set

$$\delta_{ijk} = \begin{cases} 1 \text{ if point } P_k \text{ lies on the straight line } P_i P_j, \\ 0 \text{ if point } P_k \text{ does not lie on of the straight line } P_i P_j. \end{cases}$$

Prove that the linear span of vectors $\vec{v}_{ij} = (\delta_{ij1}, \delta_{ij2}, \ldots, \delta_{ijn})$, $1 \leq i < j \leq n$, coincides with $\mathbb{R}^n$.

9. Find for which minimal $N \geq 3$ it is possible to place $N+1$ equal ellipses $E, E_1, \ldots, E_N$ on a plane in such a way that no two ellipses intersect, and for each $1 \leq i \leq N$ the ellipse E_i touches the ellipses E_{i-1}, E_{i+1}, and E (here $E_0 = E_N$ and $E_{N+1} = E_1$).

Problems for 3–4-Years Students

10. Let f be a Lebesgue measurable function on $\mathbb{R}$ such that $x^3 f(x) \to 1$, as $x \to +\infty$. Find $\lim\limits_{c \to +\infty} \left(c \cdot \int\limits_{c}^{+\infty} (x - c) f(x)\, dx \right)$.

11. Let f be a convex function on $[-1, 1]$, $0 < \alpha < 1$. Prove the inequality

$$\int_{-\pi/2}^{\pi/2} f(\alpha \sin x) dx \leq \int_{-\pi/2}^{\pi/2} f(\sin x) dx.$$

12. Assume that random vectors (ξ, η) and $(\xi, f(\xi))$ have equal distributions, where f is a Borel measurable function on $\mathbb{R}$. Prove that $\eta = f(\xi)$, almost surely.

13. See Problem 4.

14. See Problem 6.

15. Does there exist a norm on the linear space

$$X = \{x = (x_1, \ldots, x_n, 0, 0, \ldots) \mid n \geq 1,\ x_1, \ldots, x_n \in \mathbb{R}\}$$

such that the function $f(x) = x_1$, $x \in X$, is discontinuous?

16. See Problem 8.

17. See Problem 9.

18. Let $\xi_1, \ldots, \xi_n$ be independent normal random vectors in $\mathbb{R}^n$, with zero mean and unit covariance matrix. Find the expectation of the Gram determinant

$$G(\xi_1, \ldots, \xi_n) = \det\left((\xi_i, \xi_j)\right)_{i,j=1}^{n}.$$

THE PROBLEMS ARE PROPOSED BY V.B. Brayman (2, 6, 14), A.A. Dorogovtsev (5, 18), A.G. Kukush (1, 10, 12), A.G. Kukush and S.V. Shklyar (15), D. Yu. Mitin (3, 11), D.V. Radchenko (7), S.V. Slobodyanyuk (9, 17), T.D. Tymoshkevych (8, 16), and M.V. Zeldich (4, 13).

2011

Problems for 1–2-Years Students

1. Do there exist two different strictly convex functions with domain [0,1] such that their graphs intersect countably many times?

2. Let $\varphi : \mathbb{R} \to \mathbb{R}$ be an n times continuously differentiable bijective map. Prove that there exists a unique continuous function $f : \mathbb{R} \to \mathbb{R}$ for which $\varphi^{(n)} = f(\varphi)$.

3. For every n, find all symmetric $n \times n$ matrices with entries 0 and 1 such that all their eigenvalues are positive.

4. Denote by $e(P, Q, R)$ the ellipse with foci P and Q which passes through the point R. Find all the triangles ABC such that the ellipses $e(A, B, C)$, $e(B, C, A)$, and $e(C, A, B)$ share a common point.

5. Solve the equation $9^x + 4^x + 2^x = 8^x + 6^x + 1$.

6. Consider a sequence $x_n = x_{n-1} - x_{n-1}^2$, $n \geq 2$, $x_1 \in (0, 1)$. Calculate

$$\lim_{n\to\infty} \frac{n^2 x_n - n}{\ln n}.$$

7. A function $f : \mathbb{R} \to \mathbb{R}$ is called n-positive if for all n real numbers $x_1, x_2, \ldots, x_n$ satisfying $x_1 + x_2 + \ldots + x_n \geq 0$, it holds

$$\frac{1}{n}\left(f(x_1) + f(x_2) + \ldots + f(x_n)\right) \geq f(0).$$

Does there exist a function f, which is 2010-positive but not 2011-positive?

V. Brayman and A. Kukush, *Undergraduate Mathematics Competitions (1995–2016)*, Problem Books in Mathematics,
DOI 10.1007/978-3-319-58673-1_17

8. Let

$$a_n = \frac{1}{2^{n/2}\sqrt{n}} \sum_{k=0}^{n} \sqrt{\binom{n}{k}}.$$

Find $\lim\limits_{n\to\infty} a_n$.

Problems for 3–4-Years Students

9. Let ξ and η be random variables such that $\xi + \eta$ is equally distributed with ξ, and $\eta \geq 0$, almost surely. Prove that $\eta = 0$, almost surely.

10. See Problem 3.

11. Given a space X with measure λ and functions $f_n, f : X \to \mathbb{R}$, such that

$$(f_n(x))^{2011} \xrightarrow{\lambda} (f(x))^{2011} \quad \text{as } n \to \infty,$$

prove that $f_n(x) \xrightarrow{\lambda} f(x)$, as $n \to \infty$.

12. How many rings, up to isomorphism, do there exist on a set with 2011 elements?

13. Let

$$a_n = \frac{1}{\sqrt{n}} \sum_{k=0}^{n} \sqrt{\binom{n}{k} p^k (1-p)^{n-k}} \text{ for some } 0 < p < 1.$$

Find $\lim\limits_{n\to\infty} a_n$.

14. Let $x_1, \ldots, x_n \geq 0$, $\sum\limits_{i=1}^{n} x_i = 1$. Do there exist matrices

$$A_1, \ldots, A_n \in M_n(\mathbb{R})$$

such that $A_i^2 = A_i$, $A_i A_j = 0$ for $i \neq j$, and $(A_i)_{11} = x_i$?

15. Let ξ_1, ξ_2, and ξ_3 be random variables with probability density function $2x$, $x \in [0, 1]$. For all $K \geq 0$, find

$$f(K) = \min \mathsf{E}(\xi_1 + \xi_2 + \xi_3 - K)_+.$$

Here $t_+ = \max(t, 0)$, $t \in \mathbb{R}$, and the minimum is taken over all possible joint distributions of variables ξ_1, ξ_2, and ξ_3.

16. Let (X, d) be a compact metric space and T be a continuous mapping of the space X into itself. Assume that for all $x, y \in X$ the condition $\frac{1}{2}d(x, Tx) < d(x, y)$ implies that $d(Tx, Ty) < d(x, y)$. Prove that T has a unique fixed point.

The problems are proposed by V.K. Bezborodov (8, 13), A.V. Bondarenko (3, 10), V.B. Brayman (4, 5, 7, 15), M.S. Viazovska (16), A.G. Kukush (9), D. Yu. Mitin (6), O.N. Nesterenko (1), I.O. Parasyuk (2), V.M. Radchenko (11), S.V. Slobodyanyuk (12), and Ya. V. Zhurba (14).

2012

Problems for 1–2-Years Students

1. Let $a_{ij} = \tan(i - j)$, $i, j = 0, 1, \ldots, 2012$. Find $\det(a_{ij})$.

2. A function is defined on a segment, monotone, and has a primitive function. Is it necessary that the function is uniformly continuous on the segment?

3. Find the locus of incenters of triangles, for which two vertices are in the foci of a given ellipse and the third vertex lies on this ellipse.

4. For all positive numbers $x_1, \ldots, x_n$ such that

$$x_1 + \ldots + x_n \geq \max(n,\ x_1x_2 + x_2x_3 + \ldots + x_{n-1}x_n + x_nx_1)$$

prove the inequality

$$\frac{1}{x_1^{x_2}} + \frac{1}{x_2^{x_3}} + \ldots + \frac{1}{x_{n-1}^{x_n}} + \frac{1}{x_n^{x_1}} \geq \frac{n^2}{x_1^2 + \ldots + x_n^2}.$$

5. Denote by $p(n)$ the number of solutions $(x_1, \ldots, x_n)$ of the equation

$$x_1 + 2x_2 + \ldots + nx_n = n$$

in nonnegative integers. Prove that $p(n) = O(\alpha^n)$, as $n \to \infty$, for all $\alpha > 1$.

6. Find all the functions $f : \mathbb{R} \to \mathbb{R}$ such that for each $x, y \in \mathbb{R}$ it holds

$$f(x + y) + f(x - y) = 2f(x) + 2f(y),$$

and f is continuous at the point 1.

V. Brayman and A. Kukush, *Undergraduate Mathematics Competitions (1995–2016)*, Problem Books in Mathematics,
DOI 10.1007/978-3-319-58673-1_18

7. A sequence $\{a_n : n \geq 1\}$ is defined by $a_1 = 1$, $a_{n+1} = 3a_n + 2\sqrt{2(a_n^2 - 1)}$, $n \geq 1$. Prove that all its members are positive integers; moreover, the numbers a_{n+1} and a_{2n+1} are relatively prime.

8. Do there exist real 2×2 matrices A and B such that $\det A > 1$, $\det B > 1$, and for each $u_0 \in \mathbb{R}^2$ there exists a matrix sequence $\{M_i : i \geq 1\} \subset \{A, B\}$, for which the vector sequence $u_i = M_i u_{i-1}$, $i \geq 1$, is bounded?

9. Let $n \geq 3$, and $k^* = k^*(n)$ be the least number k which satisfies the inequality

$$\frac{1}{n-k} \sum_{j=k}^{n} \frac{n-j}{j} \leq 1.$$

Prove that there exists $\lim\limits_{n\to\infty} \frac{k^*(n)}{n}$.

Problems for 3–5-Years Students

10. Let (X, ρ) and (Y, σ) be metric spaces, both containing at least two points. Does there always exist a continuous function of two variables $f(x, y)$, $x \in X$, $y \in Y$, which cannot be decomposed as $g(x) + h(y)$, where g and h are some univariate functions?

11. Let ξ and η be random variables such that $\mathsf{P}(\xi \neq 0, \eta \neq 0) = 0$, and $u : \mathbb{R} \to \mathbb{R}$ be a Borel measurable bounded function. Prove that

$$\mathsf{E}u(\xi + \eta) = \mathsf{E}u(\xi) + \mathsf{E}u(\eta) - u(0).$$

12. A function $f \in C(\mathbb{R})$ has a finite limit $\lim\limits_{t\to+\infty} f(t)$. Prove that
(a) for $a > 0$, all the solutions to the equation $\dot{x} + ax = f(t)$ have the same limit, as $t \to +\infty$, and find this limit;
(b) for $a < 0$, only one of solutions to the equation has a finite limit, as $t \to +\infty$.

13. Let A and B be 3×3 complex matrices such that $A^2 = B^2 = 0$. What can be the set of eigenvalues of the matrix $A + B$?

14. Let λ be a finite measure on $(X, \mathscr{F})$, and $g_n : X \to \mathbb{R}$, $n \geq 1$ be nonnegative measurable functions such that

$$\int_X \frac{g_n^2}{1 + g_n}\, d\lambda \to 0, \quad \text{as} \quad n \to \infty.$$

Prove that $\int_X g_n\, d\lambda \to 0$, as $n \to \infty$.

15. Find all $a \in \mathbb{C}$, for which there exist vectors $v_1, v_2, \ldots$ in the complex Hilbert space l_2 such that $|v_k| = 1$ for $k \geq 1$ and $(v_i, v_j) = a$ for $i > j \geq 1$.

16. Let $f : \mathbb{R} \to \mathbb{R}$ be a function, which satisfies the equality

$$f(x+y) + f(x-y) = 2f(x) + 2f(y), \quad x, y \in \mathbb{R},$$

and f is continuous at zero point. Prove that the function

$$\varphi(x) = f(x)/x, \ x \neq 0, \ \varphi(0) = 0,$$

satisfies the equality $\varphi(x+y) = \varphi(x) + \varphi(y)$, $x, y \in \mathbb{R}$.

17. See Problem 8.

18. See Problem 9.

THE PROBLEMS ARE PROPOSED BY A.V. Bondarenko (5), S.I. Dotsenko and A.G. Kukush (9, 18), A.G. Kukush (6, 10, 11), D. Yu. Mitin (1, 4), O.N. Nesterenko (2), I.O. Parasyuk (12), V.M. Radchenko (14), M.M. Rozhkova (3), O.V. Rudenko (7), S.V. Shklyar and A.G. Kukush (16), R.V. Skuratovskyi (8, 17), and I.S. Feshchenko (13, 15).

2013

Problems for 1–2-Years Students

1. Find all continuous functions $f : [1, 2] \to [1, 2]$ such that $f(1) = 2$, and

$$f(f(x))f(x) = 2 \text{ for all } x \in [1, 2].$$

2. Does there exist a finite ring (not necessarily commutative or with identity), such that for every its element x there exists an element y different from x, for which $y^2 = x$?

3. In a given triangle, the lengths of the sides and tangents of the angles are arithmetic progressions. Find the angles of the triangle.

4. Let $x_1, \ldots, x_n$ and c be positive numbers. Prove the inequality

$$\sqrt{x_1 + \sqrt{x_2 + \sqrt{\ldots + \sqrt{x_n + c}}}} < \sqrt{x_1 + \sqrt{x_2 + \sqrt{\ldots + \sqrt{x_n}}}} + \frac{c}{2^n \sqrt{x_1 \cdot \ldots \cdot x_n}}.$$

5. Let A and B be $n \times n$ matrices such that for every $n \times n$ matrix C the equation $AX + YB = C$ has a solution X, Y. Prove that for every matrix C the equation $A^{2013}X + YB^{2013} = C$ has a solution as well.

6. Functions $f, g : \mathbb{R} \to \mathbb{R}$ are such that for every two different numbers x and y the inequality $f(x) + g(y) > 0$ or the inequality $f(y) + g(x) > 0$ holds. Prove that there are no numbers a and b such that for all $x \in (a, b)$ it holds $f(x) + g(x) < 0$.

7. A number is called *good* if it is the k-th power of a positive integer for some integer $k \geq 2$. Is the set of all positive integers which can be represented as a sum of two good numbers finite or infinite?

V. Brayman and A. Kukush, *Undergraduate Mathematics Competitions (1995–2016)*, Problem Books in Mathematics,
DOI 10.1007/978-3-319-58673-1_19

8. Let $A = \begin{pmatrix} 1 & 2 \\ 0 & 1 \end{pmatrix}$ and $B = \begin{pmatrix} 1 & 0 \\ -2 & 1 \end{pmatrix}$. Can a product $X_1X_2\dots X_n$ be equal to the identity matrix if every multiplier X_i equals either A or B?

Problems for 3–4-Years Students

9. Find the sum of the series

$$\sum_{n=0}^{\infty} \frac{1}{n!\,2^n} \cos \frac{\pi n - 1}{2}.$$

10. Does there exist a finite nonzero ring (not necessarily commutative or with identity), such that for every its nonzero element x there exists an element y different from x, for which $y^2 = x$?

11. See Problem 6.

12. Let A and B be $n \times n$ complex matrices such that for every $n \times n$ matrix C the equation $AX + YB = C$ has a solution X, Y. Prove that $k_0(A) + k_0(B) \le n$, where $k_0(U)$ is the number of zeros on the main diagonal of the Jordan form of U.

13. Let $f : \mathbb{R}^2 \to \mathbb{R}$ be a Lebesgue measurable function. Is it always true that

$$\operatorname*{ess\,sup}_{x\in\mathbb{R}} \left(\operatorname*{ess\,sup}_{y\in\mathbb{R}} f(x, y) \right) = \operatorname*{ess\,sup}_{y\in\mathbb{R}} \left(\operatorname*{ess\,sup}_{x\in\mathbb{R}} f(x, y) \right)?$$

(Here the essential suprema are taken with respect to Lebesgue measure on $\mathbb{R}$.)

14. Do there exist real nonconstant rational functions $\varphi(x)$ and $\psi(x)$ such that $\psi'(x) = \dfrac{\varphi'(x)}{\varphi(x)}$ for all x from the intersection of domains of left-hand and right-hand sides of the equality?

15. Let P be a probability measure on Borel sigma-algebra on $\mathbb{R}^2$ such that for every straight line ℓ it holds $\mathsf{P}(\ell) < 1$. Is it necessary that there exists a bounded Borel measurable set A such that for every straight line ℓ it holds $\mathsf{P}(A \cap \ell) < \mathsf{P}(A)$?

16. Let $\{a_n\}$ be a sequence of real numbers such that $\sum_{n=1}^{\infty} a_n^2 < \infty$, $\{\xi_n\}$ be a sequence of independent identically distributed random variables with distribution $\mathsf{P}\{\xi_1 = 1\} = 2/3$, $\mathsf{P}\{\xi_1 = 0\} = 1/3$, and $S_n = \sum_{k=1}^{n} \xi_k$. Prove that the series $\sum_{n=1}^{\infty} (-1)^{S_n} a_n$ converges in probability.

THE PROBLEMS ARE PROPOSED BY A.V. Bondarenko (7), I.S. Feshchenko (1, 4, 5, 12), A.G. Kukush (13, 15), A.G. Kukush and M.M. Rozhkova (3), Y.O. Makedonskyi (8, 14), D.Yu. Mitin (9), O.V. Rudenko (6, 11), G.M. Shevchenko (16), and S.V. Slobodyanyuk (2, 10).

2014

Problems for 1–2-Years Students

1. Prove that $\cos x < e^{-x^2/2}$ for all $0 < x \le \pi$.

2. It is allowed to replace a polynomial $p(x)$ with one of the polynomials $p(p(x))$, $xp(x)$, or $p(x) + x - 1$. Is it possible to transform some polynomial of the form $x^k(x-2)^{2n}$ into some polynomial of the form $x^l(x-2)^{2m+1}$ in several such steps, where k, l, m, n are positive integers?

3. Let M_A, M_B, M_C, and M_D be the intersection points of the medians of the faces BCD, ACD, ABD, and ABC of a tetrahedron $ABCD$. Points A_1 and A_2, which are symmetric with respect to M_A, are chosen on the face BCD, and points B_1 and B_2, which are symmetric with respect to M_B, are chosen on the face ACD. Prove that $V_{A_1B_1M_CM_D} = V_{A_2B_2M_CM_D}$.

4. Does there exist a real nonconstant polynomial without real roots, such that after erasing any of its terms we obtain a polynomial, which has a real root?

5. Let $x_1, x_2 \in (0, 1)$ and $(x_1, x_2) \ne \left(\frac{1}{2}, \frac{1}{2}\right)$. Prove the inequality

$$\sqrt{x_1(1-x_2)} + \sqrt{x_2(1-x_1)} \ge \sqrt{\frac{|x_1 - x_2|}{\max\left(|2x_1 - 1|, |2x_2 - 1|\right)}}.$$

When does the equality hold?

6. A square matrix P is such that $P^2 = P$, besides P is neither zero matrix nor the identity one. Does there always exist a matrix Q such that $Q^2 = Q$, $PQ = QPQ$ but $QP \ne PQ$?

V. Brayman and A. Kukush, *Undergraduate Mathematics Competitions (1995–2016)*, Problem Books in Mathematics,
DOI 10.1007/978-3-319-58673-1_20

7. Let $f, g \in C([0, 1])$, and the function f attains its maximum only at the point $x_0 \in [0, 1]$. Prove that the function

$$\varphi(t) = \max_{x\in[0,1]} (f(x) + t \cdot g(x))$$

has a derivative at zero point, and $\varphi'(0) = g(x_0)$.

Problems for 3–5-Years Students

8. Find $\int_0^\pi (\sin x)^{\cos x}\, dx$.

9. Real polynomials P_n of degree 2014 and a real polynomial P satisfy

$$\int_0^{2014} |P_n(x) - P(x)|^{2014}\, dx \to 0, \text{ as } n \to \infty.$$

Is it necessary that

$$\int_0^{2014} |P_n'(x) - P'(x)|^{2014}\, dx \to 0, \text{ as } n \to \infty?$$

10. Let $f_n : [0, 1] \times [0, 1] \to \mathbb{R}$, $n \geq 1$, be Borel measurable functions, λ_1 and λ_2 be the Lebesgue measures in $\mathbb{R}$ and $\mathbb{R}^2$, respectively. It is given that $f_n\,(x, g_n(x)) \xrightarrow{\lambda_1} 0$, as $n \to \infty$, for any sequence of Borel measurable functions $g_n : [0, 1] \to [0, 1]$. Prove that $f_n(x, y) \xrightarrow{\lambda_2} 0$, as $n \to \infty$.

11. See Problem 4.

12. See Problem 5.

13. Do there exist a sequence of independent random variables $\{\varepsilon_k,\ k \geq 3\}$ such that $\mathsf{E}\varepsilon_k = 0$, $\mathsf{Var}(\varepsilon_k) = 1$, $k \geq 3$, and a sequence of random variables $\{x_k,\ k \geq 1\}$, where

$$x_k = x_{k-1} - x_{k-2} + x_{k-3} + \varepsilon_k + \tfrac{1}{2}\varepsilon_{k-1},\ k \geq 4,$$

such that the sequence $\{\mathsf{E}x_k^2,\ k \geq 1\}$ is bounded?

14. A square matrix P is such that $P^2 = P$, besides P is neither zero matrix nor the identity one. Does there always exist a matrix Q such that $Q^2 = Q$, $PQ = QPQ$ but $QP \neq PQP$?

The problems are proposed by V.K. Bezborodov (7), V.B. Brayman (4, 11), A.G. Kukush (5, 8, 9, 12), A.G. Kukush and M.M. Rozhkova (3), D.Yu. Mitin (1, 2), V.M. Radchenko (10), O.V. Rudenko (6, 14), and S.V. Shklyar (13).

2015

Problems for 1–2-Years Students

1. Is it possible that the middle point and one of the endpoints of a segment belong to the hyperbola $y = \frac{1}{x}$, while the other endpoint of the segment belongs to the hyperbola $y = \frac{8}{x}$?

2. Solve the system of equations

$$\begin{cases} (1+\ x_1)(1+\ x_2)\dots(1+\ x_n) = 2, \\ (1+2x_1)(1+2x_2)\dots(1+2x_n) = 3, \\ \dots\quad\dots\quad\dots\quad\dots\quad\dots \\ (1+nx_1)(1+nx_2)\dots(1+nx_n) = n+1. \end{cases}$$

3. Let f be a real continuous non-monotonic function on $[0, 1]$. Prove that there exist numbers $0 \le x < y < z \le 1$ such that $z - y = y - x$ and $(f(z) - f(y))(f(y) - f(x)) < 0$.

4. Let f be a bijection on a finite set X. Prove that the number of sets $A \subseteq X$ such that $f(A) = A$ is an integer power of 2.

5. Bilbo chooses real numbers x, y and tells the numbers $x^n + y^n$ and $x^k + y^k$ to Gollum. Find all positive integers n and k for which Gollum can determine xy.

6. Let $a_0, a_1, \dots, a_n$ be real numbers such that not all of them coincide. Prove that there exists a unique solution to the equation

$$2^n \sum_{i=0}^{n} a_i e^{a_i x} = \sum_{i=0}^{n} \binom{n}{i} a_i \cdot \sum_{i=0}^{n} e^{a_i x}$$

V. Brayman and A. Kukush, *Undergraduate Mathematics Competitions (1995–2016)*, Problem Books in Mathematics,
DOI 10.1007/978-3-319-58673-1_21

7. Let A be a singular $n \times n$ matrix and B, C be column vectors of size $n \times 1$. Prove that matrix $\begin{pmatrix} A & B \\ C^{\mathrm{T}} & 0 \end{pmatrix}$ is singular if and only if $\det(A - BC^{\mathrm{T}}) = 0$.

Problems for 3–4-Years Students

8. Does the sequence of functions

$$f_n(x) = \sin^n \pi x \, \mathrm{I}_{[0,n]}(x), \; n \geq 1,$$

converge in Lebesgue measure on $\mathbb{R}$?

9. Let $f \in C(\mathbb{R}^2, \mathbb{R})$ and let $B \subset f(\mathbb{R}^2)$ be a compact set. Prove that there exists a compact set $A \subset \mathbb{R}^2$, for which $f(A) = B$.

10. Let K_C be the set of random variables distributed on $[0, 1]$, for which probability density function $p(x) \leq C$, $x \in [0, 1]$. Prove that there exists a number $a(C) > 0$ such that $\mathsf{Var}(\xi) > a(C)$ for every $\xi \in K_C$.

11. See Problem 5.

12. Let ξ be a random variable such that $\mathsf{P}(\xi > 0) > 0$ and $\mathsf{E}e^{a\xi} < \infty$ for every $a > 0$. Prove that there exists a number $\sigma > 0$ such that $\mathsf{E}e^{2\sigma\xi} = 2\mathsf{E}e^{\sigma\xi}$.

13. See Problem 7.

14. Prove that for every $x_1 < -1$, $x_2 > 1$, $y_1 \geq -1$, and $y_2 \geq -1$, the differential equation

$$xy' = \sqrt{1 + (y')^2} + y$$

has a solution $y(\cdot) \in C^{(1)}(\mathbb{R})$ such that $y(x_1) = y_1$ and $y(x_2) = y_2$.

THE PROBLEMS ARE PROPOSED BY V.B. Brayman (1, 5, 11), A.G. Kukush (12), A.G. Kukush and S.V. Shklyar (6), O.N. Nesterenko and A.V. Chaikovskiy (9), Yu.S. Mishura (10), I.O. Parasyuk (14), V.M. Radchenko (8), O.V. Rudenko (3), S.V. Shklyar (7, 13), V.G. Yurashev (2, 4).

2016

Problems for 1–2-Years Students

1. Find minimal possible value of the expression

$$4\cos^2\frac{n\pi}{9}+\sqrt[3]{7-12\cos^2\frac{n\pi}{9}},$$

where $n \in \mathbb{Z}$.

2. It is said that a set of positive integers $\{a_1 < a_2 < \ldots < a_n < \ldots\}$ has a density α if

$$\lim_{n\to\infty}\frac{\max\{k : a_k \le n\}}{n}=\alpha.$$

Does there exists a subset of positive integers with density 1 which contains no infinite increasing geometric progression?

3. Let ABC be a right isosceles triangle in $\mathbb{R}^3$, and $A_1B_1C_1$ be its orthogonal projection onto some plane. It is known that $A_1B_1C_1$ is a right isosceles triangle as well. Find all possible ratios of leg length AB to leg length A_1B_1.

4. A sequence is defined recursively:

$$x_0 = 1,\quad x_{n+1} = x_n - \frac{x_n^2}{2016},\ n \ge 0.$$

Prove that $x_{2016} < \frac{1}{2} < x_{2015}$.

V. Brayman and A. Kukush, *Undergraduate Mathematics Competitions (1995–2016)*, Problem Books in Mathematics,
DOI 10.1007/978-3-319-58673-1_22

5. Let A and B be square matrices (a) of size 2016; (b) of size 2017. Do there always exist real numbers a and b such that $a^2 + b^2 \neq 0$ and the matrix $aA + bB$ is singular?

6. Positive integers x, m, and n are such that x is divisible by 7 and $\sqrt{x} > \frac{m}{n}$. Prove that $\sqrt{x} > \dfrac{m^4 + 2m^2 + 2}{m^3 n + mn}$.

7. Let $\{x_1, x_2, \ldots, x_n\}$ and $\{y_1, y_2, \ldots, y_n\}$ be two sets of pairwise distinct real numbers, and $a_{ij} = x_i + y_j$, $1 \le i, j \le n$. It is known that the product of elements in each column of the matrix $A = (a_{ij})$ is equal to c. Find all possible products of elements in a row of the matrix A.

Problems for 3–4-Years Students

8. Some numbers are written on each face of a cubic die which falls on each edge with probability $1/12$. Is it possible that numbers on its two upper faces sum up to each of values $1, 2, \ldots, 6$ with probability $1/6$?

9. Let $f \in C^{(1)}([0, 1])$, and λ_1 be the Lebesgue measure on $\mathbb{R}$. Prove that

$$\lambda_1(\{x \in [0,1] : f(x) = 0\}) = \lambda_1\left(\{x \in [0,1] : f(x) = f'(x) = 0\}\right).$$

10. See Problem 5.

11. Let $\{A(t),\ t \in \mathbb{R}\}$ be a continuous family of skew-symmetric $n \times n$ matrices, I be the identity matrix of size n, and $X(t)$ be the solution to the matrix differential equation $\frac{dX(t)}{dt} = A(t)X(t)$, with initial value $X(0) = I$. Prove that for every point $y \in \mathbb{R}^n$, there exists a point $z \in \mathbb{R}^n$ and a sequence $\{t_i,\ i \ge 1\} \subset \mathbb{R}$ such that $t_i \to \infty$ and $X(t_i)z \to y$, as $i \to \infty$.

12. Let

$$p(x, a, b) = \begin{cases} \exp\left(ax + bx^2 + f(a,b) + g(x)\right), & x \in [0; 1], \\ 0, & x \in \mathbb{R} \setminus [0; 1], \end{cases}$$

be a family of probability density functions with parameters $a, b \in \mathbb{R}$, and $g \in C(\mathbb{R})$. Prove that

$$\left(f'_a(a,b)\right)^2 + f'_b(a,b) < 0, \quad a, b \in \mathbb{R}.$$

13. Let $K : [0, 1] \to [0, 1]$ be the Cantor's function, i.e., K is the nondecreasing function such that $K\left(\sum_{i \in S} \frac{2}{3^i}\right) = \sum_{i \in S} \frac{1}{2^i}$, for every set $S \subset \mathbb{N}$. Find

$$\lim_{n \to \infty} n \int_{[0,1]} K^n(x)\, d\lambda_1.$$

14. Let A be a real matrix of size $m \times n$, $m < n$, such that $\operatorname{rk} A = m$, and I be the identity matrix of size m. Construct a real $n \times m$ matrix X with the least possible sum of squared entries, for which $AX = I$.

THE PROBLEMS ARE PROPOSED BY A. V. Bondarenko (5, 10), V. B. Brayman (3, 8), D. I. Khilko (2), A. G. Kukush (12, 14), I. O. Parasyuk (11), V. M. Radchenko (13), O. V. Rudenko (9), D. V. Tkachenko (4, 7), and V. G. Yurashev (1, 6).

Part II
Solutions

Other mistakes may perchance, having eluded us both, await the penetrating glance of some critical reader, to whom the joy of discovery, and the intellectual superiority which he will thus discern, in himself, to the author of this little book, will, I hope, repay to some extent the time and trouble its perusal may have cost him!

Charles L. Dodgson, "*Pillow-Problems thought out during wakeful hours*"

1995

1 The function $f(t) = (t-1)\ln t$ is continuous and strictly increasing on the interval $[1, +\infty)$, moreover $f(1) = 0$ and $f(t) \to +\infty$, as $t \to +\infty$. By the intermediate value theorem, f attains all positive integers, i.e., for each $n \in \mathbb{N}$ there exists $t(n) > 1$ such that $f(t) = n$, and $t(n)$ is unique due to the monotonicity of f. Since $\frac{\ln n}{n} \to 0$, as $n \to \infty$, we have

$$\lim_{n\to\infty} t(n)\frac{\ln n}{n} = \lim_{n\to\infty} (t(n)-1)\frac{\ln n}{n} = \lim_{n\to\infty} \frac{\ln n}{\ln t(n)} =$$
$$= \lim_{t\to+\infty} \frac{\ln f(t)}{\ln t} = \lim_{t\to+\infty} \frac{\ln(t-1) + \ln\ln t}{\ln t} = 1,$$

because $n = f(t(n))$, $n \in \mathbb{N}$ and $t(n) \to +\infty$, as $n \to \infty$.

2 We will show that if $\{a_n, n \ge 1\}$ is bounded then B is either a point or a segment. Then the equality $A = B$ will imply the statement of the problem.

Assume that B is not a point. Then $\alpha = \liminf\limits_{n\to\infty} b_n < \beta = \limsup\limits_{n\to\infty} b_n$. If additionally B is not a segment then there exists a number γ, for which $\alpha < \gamma < \beta$ and $\gamma \notin B$. Then there exists $\varepsilon > 0$ such that $(\gamma - \varepsilon, \gamma + \varepsilon) \subset [\alpha, \beta]$, and the interval $(\gamma - \varepsilon, \gamma + \varepsilon)$ contains only finite number of elements of the sequence $\{b_n, n \ge 1\}$. Let $\{a_n, n \ge 1\} \subset [-c, c]$. Choose n_0 such that for each $n \ge n_0$ the number b_n does not belong to $(\gamma - \varepsilon, \gamma + \varepsilon)$ and the inequalities $\frac{n(\gamma-\varepsilon)+c}{n+1} < \gamma$, $\frac{n(\gamma+\varepsilon)-c}{n+1} > \gamma$ hold (such n_0 exists because $\frac{n(\gamma-\varepsilon)+c}{n+1} \to \gamma - \varepsilon < \gamma$ and $\frac{n(\gamma+\varepsilon)-c}{n+1} \to \gamma + \varepsilon > \gamma$, as $n \to \infty$). Assume that $n \ge n_0$ and $b_n \le \gamma - \varepsilon$. Then

$$b_{n+1} = \frac{nb_n + a_{n+1}}{n+1} \le \frac{n(\gamma-\varepsilon)+c}{n+1} < \gamma,$$

V. Brayman and A. Kukush, *Undergraduate Mathematics Competitions (1995–2016)*, Problem Books in Mathematics,
DOI 10.1007/978-3-319-58673-1_23

Therefore, $b_{n+1} \le \gamma - \varepsilon$, because $b_{n+1} \notin (\gamma - \varepsilon, \gamma + \varepsilon)$. Similarly $b_n \ge \gamma + \varepsilon$ implies that $b_{n+1} \ge \gamma + \varepsilon$. Thus, either $b_n \le \gamma - \varepsilon$ for all $n \ge n_0$, or $b_n \ge \gamma + \varepsilon$ for all $n \ge n_0$, hence α and β cannot both be limit points of $\{b_n, n \ge 1\}$, a contradiction. So, $B = [\alpha, \beta]$.

Let $\{a_1, a_2, a_3, a_4, \dots\} = \{0, q_1, 0, q_2, \dots\}$, where $q_1, q_2, \dots$ are all the rational numbers of the segment [0, 1] enumerated in some order. Then it is clear that $A = [0, 1]$, and because of $0 \le b_n \le \frac{1}{2}$ for all $n \ge 1$, we have $B \subset [0, \frac{1}{2}]$, in particular $B \ne A$.

3 We have $2xF(x) - F'(x) = 0$,

$$e^{-x^2}\left(2xF(x) - F'(x)\right) = -\left(e^{-x^2}F(x)\right)' = 0.$$

Thus, $e^{-x^2}F(x) = \text{const}$, $F(x) = ce^{x^2}$, $x \in \mathbb{R}$, where $c \in \mathbb{R}$ is some constant. Hence $f(x) = 2cxe^{x^2} = c_1xe^{x^2}$.
Answer: $f(x) = cxe^{x^2}$, where $c \in \mathbb{R}$ is an arbitrary constant.

4 Introduce a function $g(x) = (1 - x)\int_0^x f(t)dt$, $x \in [0, 1]$. Since $g \in C^{(1)}([0, 1])$, $g(0) = g(1) = 0$, by Rolle's Theorem

$$\exists c \in (0, 1): \quad g'(c) = -\int_0^c f(t)dt + (1 - c)f(c) = 0.$$

5 The matrix $A_0 = A$ is symmetric and positive definite, therefore, it has an eigenbasis and all its eigenvalues λ_i, $1 \le i \le m$, are positive. We have $\sum_{i=1}^{m} \lambda_i = \operatorname{tr} A < 1$, so $0 < \lambda_i < 1$, $1 \le i \le m$. It holds $A_{n+1} - \frac{1}{2}I = \left(A_n - \frac{1}{2}I\right)^2 = \left(A - \frac{1}{2}I\right)^{2^{n+1}}$. Therefore, all the matrices have common eigenbasis, and moreover the eigenvalues of matrix A_n are equal to $\frac{1}{2} + \left(\lambda_i - \frac{1}{2}\right)^{2^n} \to \frac{1}{2}$, as $n \to \infty$, because $\left|\lambda_i - \frac{1}{2}\right| < \frac{1}{2}$, $1 \le i \le m$. Thus, $A_n \to \frac{1}{2}I$, as $n \to \infty$.

6 Notice that $|\sin x - \sin a| \le |x - a|$, $x \in \mathbb{R}$, and use the QM-AM inequality to get

$$\left|\frac{1}{n}\sum_{k=1}^{n} \sin x_k - \sin a\right| \le \frac{1}{n}\sum_{k=1}^{n} |\sin x_k - \sin a| \le \frac{1}{n}\sum_{k=1}^{n} |x_k - a| \le$$
$$\le \left(\frac{1}{n}\sum_{k=1}^{n} |x_k - a|^2\right)^{1/2} = \left(\frac{1}{n}\sum_{k=1}^{n} x_k^2 - \frac{2}{n}\sum_{k=1}^{n} x_k \cdot a + a^2\right)^{1/2} \to 0, \text{ as } n \to \infty.$$

7 It is evident that a plane can be tiled with equal quadrangles, which are similar to F and have disjoint interiors (see Fig. 1).

We tile a plane with such quadrangles of area $\frac{1}{n}$. The diameters of the quadrangles equal to $\frac{1}{\sqrt{n}}d(F)$, where $d(F)$ stands for the diameter of F. Therefore, the quadrangles which lie within the disc G completely cover the concentric to G disc of radius $\frac{1}{\sqrt{\pi}} - \frac{1}{\sqrt{n}}d(F)$. Thus, their total area is not less than

$$\pi\left(\frac{1}{\sqrt{\pi}} - \frac{1}{\sqrt{n}}d(F)\right)^2 = 1 - 2\sqrt{\frac{\pi}{n}}d(F) + \frac{\pi}{n}d^2(F) \to 1, \text{ as } n \to \infty.$$

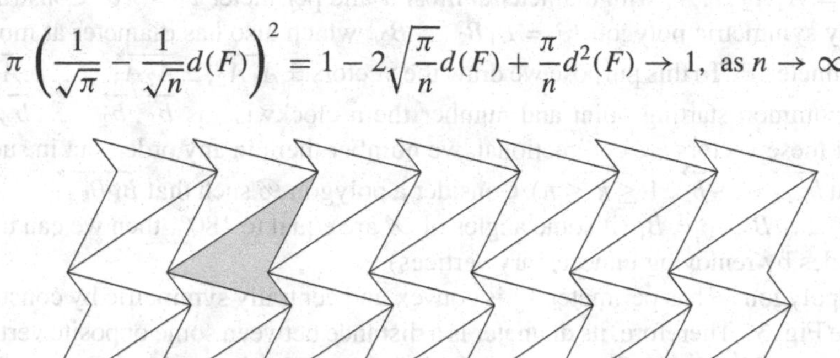

Fig. 1 A plane tiled with equal quadrangles

On the other hand, one cannot place more than n quadrangles of area $\frac{1}{n}$ with disjoint interiors in a disc of unit area. Thus, $\frac{a(n)}{n} \leq 1$, $\lim\limits_{n\to\infty} \frac{a(n)}{n} = 1$.

Consider $b(n)$ discs of area $\frac{1}{n}$ which lie inside F and have disjoint interiors. Without loss of generality we may assume that each circle touches at least one another circle (we can shift some circle otherwise). For each circle ω, consider equilateral triangle shown in Fig. 2 (ω' is one of the circles which touch ω).

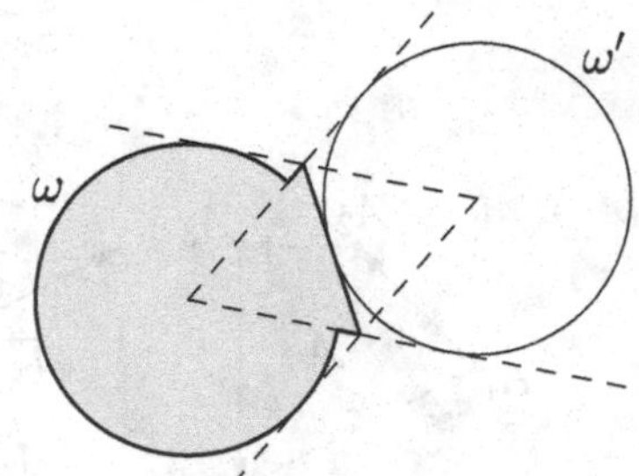

Fig. 2 The shaded figure is a union of a disc and an equilateral triangle

It is evident that the figures of area $\frac{1}{n}\left(1 + \left(\frac{1}{\sqrt{3\pi}} - \frac{1}{6}\right)\right)$, each of which is a union of a disc and the corresponding triangle, have disjoint interiors. Their total area $\frac{b(n)}{n}\left(1 + \frac{1}{\sqrt{3\pi}} - \frac{1}{6}\right)$ does not exceed the area of F, that is 1, whence

$$\limsup_{n\to\infty} \frac{b(n)}{n} \leq \left(1 + \frac{1}{\sqrt{3\pi}} - \frac{1}{6}\right)^{-1} < 1.$$

8 The length of circle with diameter d equals πd. Prove that it is the maximum.

Assume that some convex piecewise-smooth (hence rectifiable) contour with diameter d has perimeter larger than πd. Inscribe into the contour a convex polygon $\mathscr{A} = A_1A_2\dots A_n$ with diameter at most d and perimeter $P > \pi d$. Construct a centrally symmetric polygon $\mathscr{B} = B_1B_2\dots B_{2n}$ which also has diameter at most d and perimeter P. To this purpose we draw the vectors $\pm\overrightarrow{A_1A_2}, \pm\overrightarrow{A_2A_3}, \dots, \pm\overrightarrow{A_nA_1}$ with a common starting point and number them clockwise as $\overrightarrow{b_1}, \overrightarrow{b_2}, \dots, \overrightarrow{b_{2n}}$ (if some of these vectors are codirectional, we number them in any order, but in such a way that $\overrightarrow{b_{n+k}} = -\overrightarrow{b_k}$, $1 \le k \le n$). Consider a polygon $\mathscr{B}$ such that $\overrightarrow{B_kB_{k+1}} = \frac{1}{2}\overrightarrow{b_k}$, $1 \le k \le 2n$, $B_{2n+1} = B_1$ (if some angles of $\mathscr{B}$ are equal to 180°, then we can unify some sides by removing unnecessary vertices).

The polygon $\mathscr{B}$ has perimeter P, is convex and centrally symmetric by construction (see Fig. 3). Therefore, its diameter is a distance between some opposite vertices B_k and B_{n+k}. Hence, the diameter equals the sum of absolute values of projections of vectors $\overrightarrow{b_k}, \dots, \overrightarrow{b_{n+k-1}}$ onto $\overrightarrow{B_kB_{n+k}}$, i.e., it equals a half of the sum of absolute values of projections of vectors $\overrightarrow{A_1A_2}, \dots, \overrightarrow{A_nA_1}$ onto $\overrightarrow{B_kB_{n+k}}$. So it equals the projection of $\mathscr{A}$ on $\overrightarrow{B_kB_{n+k}}$ and does not exceed the diameter of $\mathscr{A}$. Thus, the diameter of $\mathscr{B}$ does not exceed d. Denote by O the center of the polygon $\mathscr{B}$. Then $\mathscr{B}$ is contained in a disc with center O and radius $\frac{d}{2}$. Consider a polygon $\mathscr{C} = C_1C_2\dots C_{2n}$ such that B_k lies on OC_k and $OC_k = \frac{d}{2}$, $1 \le k \le 2n$. The perimeter of $\mathscr{C}$ is less than πd, because each side of $\mathscr{C}$ is less than the corresponding arc. On the other hand, the perimeter of $\mathscr{C}$ is not less then the perimeter of $\mathscr{B}$ (to prove this one can consecutively cut off the parts of polygon $\mathscr{C}$, which lie outside $\mathscr{B}$, along the lines B_kB_{k+1} and apply the polygon inequality), hence the perimeter of $\mathscr{C}$ is greater than πd. We get a contradiction, so indeed πd is the maximum length of a contour with diameter d.

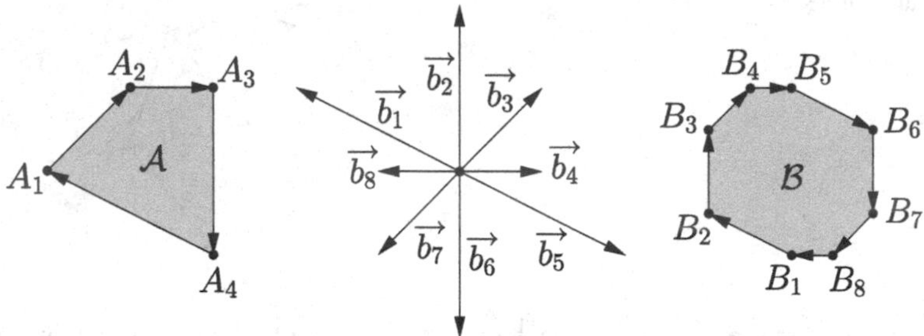

Fig. 3 From a convex polygon $\mathscr{A}$ to a centrally symmetric convex polygon $\mathscr{B}$

Remark 1 Maximum is attained not only for a circle, but also for other *constant width curves*, i.e., convex contours, for which the length of projection on any straight line equals d. An example of constant width curve is Reuleaux triangle, which consists of three arcs of a circle with radius d subtending an angle 60°.

9 By the formula of solution to linear differential equation we have

$$y(x) = \left(c + \int_0^x (\arctan t) e^{-2t-\sin t}\, dt \right) \cdot e^{2x+\sin x}.$$

Improper integral $\int_0^\infty (\arctan t) e^{-2t-\sin t} dt$ converges, because

$$(\arctan t) e^{-2t-\sin t} \le \tfrac{\pi}{2} e^{-2t+1}.$$

It holds $e^{2x+\sin x} \to +\infty$, as $x \to +\infty$. Therefore, for each

$$c \ne -\int_0^\infty (\arctan t) e^{-2t-\sin t} dt$$

the solution is unbounded. It suffices to show that the solution

$$y(x) = -e^{2x+\sin x} \int_x^\infty (\arctan t) e^{-2t-\sin t} dt$$

is bounded on $\mathbb{R}$. We have

$$|y(x)| \le e^{2x+1} \int_x^\infty \frac{\pi}{2} e^{-2t+1} dt = e^{2x+1} \cdot \frac{\pi e}{4} e^{-2x} = \frac{\pi e^2}{4}, \quad x \in \mathbb{R}.$$

10 It is easy to see that $y(x) = x$ is a solution to the Cauchy problem. Prove that the solution is unique. Suppose that $z(x)$ is another solution. Then

$$\begin{aligned} |y'(x) - z'(x)| &= \left| \int_0^x (\sin(y(u)) - \sin(z(u)))\, du \right| \le \\ &\le \int_0^x |\sin(y(u)) - \sin(z(u))|\, du \le \int_0^x |y(u) - z(u)|\, du. \end{aligned}$$

If $t = \inf\{x > 0 : y(x) \ne z(x)\} < \infty$, then $0 < c = \sup\limits_{x \le t+1} |y(x) - z(x)| < \infty$. But for $x \le t$ it holds $y(x) = z(x)$, and for $t \le x \le t+1$ we have

$$\begin{aligned} |y(x) - z(x)| &= \left| \int_t^x \left(y'(s) - z'(s) \right) ds \right| \le \int_t^x \int_0^s |y(u) - z(u)| du\, ds = \\ &= \int_t^x \int_t^s |y(u) - z(u)| du\, ds \le \frac{c(x-t)^2}{2} \le \frac{c}{2}. \end{aligned}$$

Thus, $c \le \frac{c}{2}$, a contradiction. So $y(x) = x$ is the unique solution.

11 Set $z_1 = e^{\frac{2\pi i}{k}}$, $z_j = z_1^j$, $j = 0, \ldots, k-1$. We have

$$\sum_{j=0}^{k-1} z_j^{k-l} f(z \cdot z_j) = \sum_{n=0}^{\infty} c_n z^n \cdot \sum_{j=0}^{k-1} z_j^{n+k-l} = \sum_{m=0}^{\infty} c_{km+l} z^{km+l} \cdot k = 0,$$

because

$$\sum_{j=0}^{k-1} z_j^{n+k-l} = \begin{cases} k, \text{ if } n = km + l, \\ 0 \text{ otherwise.} \end{cases}$$

Thus, $\sum\limits_{j=0}^{k-1} z_j^{k-l} f(z \cdot z_j) = 0$ for all z such that $|z| < 1$, hence for $|z| = 1$. Since the radius of convergence of the series equals 1, then there exists at least one singular point of f on the unit circle (otherwise f has an analytic extension on a neighborhood of each point of the unit circle, therefore, the radius of convergence exceeds 1).

Suppose that z_* is a unique singular point of f on the unit circle. Then z_* is a singular point of the function $\sum\limits_{j=0}^{k-1} z_j^{k-l} f(z \cdot z_j) = 0$ as well, because it is a singular point of the function $z_j^{k-l} f(z \cdot z_j)$ if and only if $j = 0$, a contradiction. Thus, there exists a number $j \neq 0$ such that $z_* z_j$ also is a singular point of f.

12 Introduce polar coordinates (r, φ) and let

$$\overline{u}(r, \varphi) = u(r \cos \varphi, r \sin \varphi).$$

The set W corresponds to a set $\overline{W}$ of functions $\overline{u}(r, \varphi)$ such that $r^2 \frac{\partial^2 \overline{u}}{\partial r^2} + \frac{\partial^2 \overline{u}}{\partial \varphi^2} = 0$ in a ring $\overline{K} = \{(r, \varphi) \mid 1 \le r \le 2,\ 0 \le \varphi < 2\pi\}$ and $\int_0^{2\pi} j \frac{\partial \overline{u}}{\partial r}(j, \varphi) d\varphi = (-1)^{j+1} 2\pi$, $j = 1, 2$. In polar coordinates we have

$$\overline{D}(\overline{u}) = \int_1^2 \int_0^{2\pi} \left(r \left(\tfrac{\partial \overline{u}}{\partial r}(r, \varphi) \right)^2 + \tfrac{1}{r} \left(\tfrac{\partial \overline{u}}{\partial \varphi}(r, \varphi) \right)^2 \right) d\varphi dr.$$

For each function $\overline{u} \in \overline{W}$, it is easy to check that a function $\overline{v}(r, \psi) = \frac{1}{2\pi} \int_0^{2\pi} \overline{u}(r, \varphi)$ $d\varphi$ belongs to $\overline{W}$ as well. Since

$$\tfrac{\partial \overline{v}}{\partial r}(r, \psi) = \tfrac{1}{2\pi} \int_0^{2\pi} \tfrac{\partial \overline{u}}{\partial r}(r, \varphi) d\varphi, \quad \tfrac{\partial \overline{v}}{\partial \psi}(r, \psi) = 0,$$

by the Cauchy–Schwarz inequality it holds

$$\int_0^{2\pi} \left(\tfrac{\partial \overline{v}}{\partial r}(r, \psi) \right)^2 d\psi = \int_0^{2\pi} \left(\tfrac{1}{2\pi} \int_0^{2\pi} \tfrac{\partial \overline{u}}{\partial r}(r, \varphi) d\varphi \right)^2 d\psi =$$

$$= 2\pi \left(\tfrac{1}{2\pi} \int_0^{2\pi} \tfrac{\partial \overline{u}}{\partial r}(r, \varphi) d\varphi \right)^2 \le \int_0^{2\pi} \left(\tfrac{\partial \overline{u}}{\partial r}(r, \varphi) \right)^2 d\varphi \text{ for each } r \in [1, 2].$$

Therefore, $\overline{D}(\overline{v}) \le \overline{D}(\overline{u})$, and in the case $\int_1^2 \int_0^{2\pi} \frac{1}{r} \left(\frac{\partial \overline{u}}{\partial \varphi}(r, \varphi) \right)^2 d\varphi dr \ne 0$ the inequality is strict. Thus, if $\overline{D}(\overline{u}^*) = \min\limits_{\overline{u} \in \overline{W}} \overline{D}(\overline{u})$, then the function $\overline{u}^*$ satisfies the condition $\frac{\partial \overline{u}^*}{\partial \varphi}(r, \varphi) = 0$, $(r, \varphi) \in \overline{K}$, that is $\overline{u}^*$ does not depend on φ. In particular the corresponding function u^* is constant on both S_1 and S_2.

13 We assume that the hare continues jumping after it has been trapped. The sizes of the jumps are jointly independent, and the size of a jump is positive with probability $\frac{2}{3}$. Therefore, by the Borel–Cantelli lemma the hare will almost surely make infinitely many positive jumps. Thus, with probability 1 it will visit points with arbitrary large coordinates. Maximum length of jump equals 2. Hence, the hare will almost surely visit all the sets

$$\{1, 2\}, \{3, 4\}, \ldots, \{2n - 1, 2n\}, \ldots$$

Both numbers in couple are traps with probability 0.16, so with probability 1 there exists a couple $\{2n - 1, 2n\}$ which contains two traps. Therefore, the hare will be trapped almost surely.

14 Since

$$\forall x \in H \ \exists c_x > 0 \ \forall n \ge 1 \ \|A_n x\| \le c_x,$$

by Banach–Steinhaus Theorem there exists $C > 0$ such that $\|A_n\| \le C$ for all $n \ge 1$. We have

$$\|A_n K\| = \sup_{\|x\|=1} \|A_n K x\| = \sup\{\|A_n y\| \mid y = Kx, \ \|x\| = 1\}.$$

A set $\{y = Kx, \ \|x\| = 1\}$ is compact, thus, for every $\varepsilon > 0$ there exists a finite ε-net $\{y_1, \ldots, y_m\}$. Hence

$$\|A_n y\| \le \|A_n y_k\| + \|A_n(y - y_k)\| \le \|A_n y_k\| + C\|y - y_k\| \le \|A_n y_k\| + C\varepsilon,$$

where y_k is an element of the net, for which $\|y - y_k\| < \varepsilon$. Therefore,

$$\|A_n K\| \le \max_{1 \le k \le m} \|A_n y_k\| + C\varepsilon \to C\varepsilon, \text{ as } n \to \infty.$$

Since $\varepsilon > 0$ is arbitrary, it holds $\|A_n K\| \to 0$, as $n \to \infty$.

1996

1 Let $a = a_0 e^{i\alpha}$, $b = b_0 e^{i\beta}$, and $c = c_0 e^{i\gamma}$, where $a_0, b_0, c_0 \geq 0$ and $\alpha, \beta, \gamma \in [0, 2\pi)$. Without loss of generality we may assume that $a_0 \geq b_0 \geq c_0$. Then

$$\limsup_{n\to\infty} \left|a^n + b^n + c^n\right|^{\frac{1}{n}} \leq \limsup_{n\to\infty} \left(|a|^n + |b|^n + |c|^n\right)^{\frac{1}{n}} \leq \limsup_{n\to\infty} (3a_0^n)^{\frac{1}{n}} = a_0.$$

Construct an increasing sequence of numbers $\{n_k, k \geq 1\}$ such that

$$\left|a^{n_k} + b^{n_k} + c^{n_k}\right|^{\frac{1}{n_k}} \geq a_0,\ k \geq 1.$$

By the pigeonhole principle, there exist infinitely many numbers m_k such that all the numbers $e^{im_k(\beta-\alpha)}$ lie in the same coordinate quadrant and all the numbers $e^{im_k(\gamma-\alpha)}$ lie in the same (possibly other) coordinate quadrant as well. Therefore, for $n_k = m_{k+1} - m_1$, $k \geq 1$, it holds $\operatorname{Re} e^{in_k(\beta-\alpha)} \geq 0$ and $\operatorname{Re} e^{in_k(\gamma-\alpha)} \geq 0$. Hence

$$\begin{aligned}\left|a^{n_k} + b^{n_k} + c^{n_k}\right| = \left|a_0^{n_k} + b_0^{n_k} e^{in_k(\beta-\alpha)} + c_0^{n_k} e^{in_k(\gamma-\alpha)}\right| \geq \\ \geq a_0^{n_k} + b_0^{n_k} \operatorname{Re} e^{in_k(\beta-\alpha)} + c_0^{n_k} \operatorname{Re} e^{in_k(\gamma-\alpha)} \geq a_0^{n_k}.\end{aligned}$$

Thus, $\limsup_{n\to\infty} |a^n + b^n + c^n|^{\frac{1}{n}} \geq a_0$, so

$$\limsup_{n\to\infty} \left|a^n + b^n + c^n\right|^{\frac{1}{n}} = a_0 = \max\{|a|, |b|, |c|\}.$$

Answer: $\limsup_{n\to\infty} |a^n + b^n + c^n|^{\frac{1}{n}} = \max\{|a|, |b|, |c|\}$.

V. Brayman and A. Kukush, *Undergraduate Mathematics Competitions (1995–2016)*, Problem Books in Mathematics,
DOI 10.1007/978-3-319-58673-1_24

2 For $x, y \geq 1$, it holds

$$\varphi(xy) - \varphi(x) = \lim_{A\to\infty} \left(\int_A^{Axy} f(u)du - \int_A^{Ax} f(u)du \right) =$$
$$= \lim_{A\to\infty} \int_{Ax}^{Axy} f(u)du = \lim_{Ax\to\infty} \int_{Ax}^{Axy} f(u)du = \varphi(y),$$

that is $\varphi(xy) = \varphi(x) + \varphi(y)$. Notice that for each $c \in \mathbb{R}$ the function $\varphi(x) = c \ln x$ satisfies the functional equation

$$\varphi(xy) = \varphi(x) + \varphi(y), \; x, y \geq 1.$$

Show that this equation has no other solution which is continuous at point 1. Let $\varphi(x)$ be any such solution, $c = \varphi(e)$, and $\psi(x) = \varphi(x) - c \ln x$, $x \geq 1$. Then ψ is other solution, and $\psi(1) = \psi(e) = 0$. It is easy to prove by induction on n that $\psi(x^n) = n\psi(x)$, thus,

$$\forall n, m \geq 1 \;\; \psi\left(x^{\frac{n}{m}}\right) = n\psi\left(x^{\frac{1}{m}}\right) = \tfrac{n}{m}\psi(x),$$

hence $\psi(e^r) = 0$, $r \in \mathbb{Q}$, $r \geq 0$. Assume that $\psi(a) \neq 0$ for some $a > 1$. Since the set of numbers e^r, $r \in \mathbb{Q}$, is dense in $[1, +\infty)$, for each $n \geq 1$ there exists $r_n \in \mathbb{Q}$ such that $1 < \frac{a}{e^{r_n}} < 1 + \frac{1}{n}$, so $\frac{a}{e^{r_n}} \to 1$, as $n \to \infty$. But

$$\psi(a) = \psi\left(e^{r_n}\right) + \psi\left(\frac{a}{e^{r_n}}\right) = \psi\left(\frac{a}{e^{r_n}}\right) \not\to 0 = \psi(1), \; \text{as } n \to \infty,$$

a contradiction with continuity of ψ at point 1. Therefore, $\psi(x) \equiv 0$, $x \geq 1$, i.e. $\varphi(x) = c \ln x$, $x \geq 1$. For $f(x) = \frac{c}{x}$, $x \geq 1$ and $A \geq 1$ we have an identity

$$\int_A^{Ax} f(u)du = c \ln x.$$

Thus, the function $\varphi(x) = c \ln x$ satisfies the conditions of the problem for each $c \in \mathbb{R}$.
Answer: $\varphi(x) = c \ln x$, $c \in \mathbb{R}$.

3 Introduce a function $F(x) = \int_0^x f^2(u)du$. It holds $F'(x) = f^2(x)$, and condition of the problem takes a form $f(x)F(x) \to 1$, as $x \to +\infty$, or $F'(x)F^2(x) \to 1$, as $x \to +\infty$. For every $\varepsilon > 0$ there exists $N > 0$ such that $|F'(x)F^2(x) - 1| < \varepsilon$, $x \geq N$. Then for all $x \geq N$ it holds

$$F^3(N) + 3(x - N)(1 - \varepsilon) \leq F^3(x) =$$
$$= F^3(N) + \int_N^x 3F'(x)F^2(x)\,dx \leq F^3(N) + 3(x - N)(1 + \varepsilon),$$

hence $\dfrac{F^3(x)}{3x} \to 1$, as $x \to +\infty$, or $\dfrac{F(x)}{(3x)^{\frac{1}{3}}} \to 1$, as $x \to +\infty$. Finally,

$$f(x) \sim \frac{1}{F(x)} \sim \frac{1}{(3x)^{\frac{1}{3}}}, \text{ as } x \to +\infty.$$

4 Consider an arbitrary partition $\lambda = \{0 = x_0 < x_1 < \ldots < x_{n-1} < x_n = 1\}$. Since

$$\sum_{k=0}^{n-1} \int_{x_k}^{x_{k+1}} \sin 2\pi t \, dt = \int_0^1 \sin 2\pi t \, dt = 0,$$

it holds

$$\sum_{k=0}^{n-1} (x_{k+1} - x_k) \sin 2\pi x_k = \sum_{k=0}^{n-1} \left((x_{k+1} - x_k) \sin 2\pi x_k - \int_{x_k}^{x_{k+1}} \sin 2\pi t \, dt \right) =$$

$$= \sum_{k=0}^{n-1} \int_{x_k}^{x_{k+1}} (\sin 2\pi x_k - \sin 2\pi t) dt = -2\pi \sum_{k=0}^{n-1} \int_{x_k}^{x_{k+1}} (x_{k+1} - t) \cos 2\pi t \, dt =$$

$$= -2\pi \sum_{k=0}^{n-1} \int_{x_k}^{x_{k+1}} (x_{k+1} - t) \, dt \cdot \cos 2\pi \theta_k = -\pi \sum_{k=0}^{n-1} (x_{k+1} - x_k)^2 \cdot \cos 2\pi \theta_k$$

for some numbers $\theta_k \in [x_k, x_{k+1}]$ (we applied integration by parts and the intermediate value theorem).

Hence for every partition λ it holds

$$\sum_{k=0}^{n-1} (x_{k+1} - x_k) \sin 2\pi x_k \le \pi \left(\sum_{k=0}^{n-1} (x_{k+1} - x_k)^2 \right).$$

On the other hand, consider the partition

$$\lambda_N = \left\{ 0 < \frac{1}{2N^3} < \ldots < \frac{N^3}{2N^3} < \frac{N^3 + N^2}{2N^3} < \ldots < \frac{2N^3 - 1}{2N^3} < 1 \right\},$$

which has $2N^3 - N^2$ sub-intervals of length $\frac{1}{2N^3}$ and sub-interval $[\frac{1}{2}, \frac{1}{2} + \frac{1}{2N}]$. For this partition, we have

$$\frac{\sum_{k=0}^{n-1} (x_{k+1} - x_k) \sin 2\pi x_k}{\sum_{k=0}^{n-1} (x_{k+1} - x_k)^2} = \frac{-\pi \sum_{k=0}^{n-1} (x_{k+1} - x_k)^2 \cdot \cos 2\pi \theta_k}{\sum_{k=0}^{n-1} (x_{k+1} - x_k)^2} >$$

$$> \frac{\pi\left(\cos\frac{\pi}{N}\cdot\frac{1}{4N^2}-\frac{2N^3-N^2}{4N^6}\right)}{\frac{1}{4N^2}+\frac{2N^3-N^2}{4N^6}} \to \pi, \text{ as } N\to\infty.$$

Thus, the supremum in question equals π.
Answer: π.

5 By Riemann–Schwarz symmetry principle, the function

$$F(z)=\begin{cases} f(z), & \text{if } \operatorname{Im} z\ge 0,\\ \overline{f(\bar z)}, & \text{if } \operatorname{Im} z<0\end{cases}$$

is analytic on $\mathbb{C}\backslash\{\pm i\}$. Let $\lim\limits_{z\to i} f(z)(z-i)=c$. Then the function

$$G(z)=F(z)-\frac{c}{z-i}-\frac{\bar c}{z+i}$$

is analytic on $\mathbb{C}$, and $\lim\limits_{z\to\infty} G(z)=A$. Hence, by Liouville's theorem $G(z)\equiv A$, $z\in\mathbb{C}$, that is $f(z)=A+\frac{c}{z-i}+\frac{\bar c}{z+i}$, $\operatorname{Im} z\ge 0$.

6 Since $\operatorname{Im}\left(\frac{f(z)}{F(z)}+1\right)=\operatorname{Im}\frac{f(z)+F(z)}{F(z)}\neq 0$ for $z\in\partial D$, we have $f(z)+F(z)\neq 0$, $z\in\partial D$. Therefore, the number of roots of $F(z)$ in D equals $\frac{1}{2\pi i}\int_{\partial D}\frac{F'(z)}{F(z)}\,dz$, and the number of roots of $f(z)+F(z)$ in D equals $\frac{1}{2\pi i}\int_{\partial D}\frac{f'(z)+F'(z)}{f(z)+F(z)}\,dz$, respectively. We have

$$\frac{1}{2\pi i}\int_{\partial D}\frac{f'(z)+F'(z)}{f(z)+F(z)}\,dz=\frac{1}{2\pi i}\int_{\partial D}\frac{d}{dz}\ln\left(f(z)+F(z)\right)dz=$$
$$=\frac{1}{2\pi i}\int_{\partial D}\frac{d}{dz}\ln F(z)dz+\frac{1}{2\pi i}\int_{\partial D}\frac{d}{dz}\ln\left(1+\frac{f(z)}{F(z)}\right)dz,$$

$$\frac{1}{2\pi i}\int_{\partial D}\frac{F'(z)}{F(z)}dz=\frac{1}{2\pi i}\int_{\partial D}\frac{d}{dz}\ln F(z)\,dz.$$

It remains to prove that

$$\frac{1}{2\pi i}\int_{\partial D}\frac{d}{dz}\ln\left(1+\frac{f(z)}{F(z)}\right)dz=0.$$

Let $\omega(z)=1+\frac{f(z)}{F(z)}$ traverse the contour γ, as z traverses ∂D. Then it holds either $\gamma\subset\{\operatorname{Im}\omega>0\}$ or $\gamma\subset\{\operatorname{Im}\omega<0\}$, because $\operatorname{Im}\left(1+\frac{f(z)}{F(z)}\right)\neq 0$. Hence

$$\int_{\partial D} \frac{d}{dz} \ln \left(1 + \frac{f(z)}{F(z)}\right) dz = \int_{\gamma} \frac{d\omega}{\omega} = 0.$$

7 The eigenvalues of the matrix A are the roots of the polynomial

$$P_A(\lambda) = \lambda^{1996} + \lambda^{998} + 1996.$$

Since

$$P_A'(\lambda) = 1996\lambda^{1995} + 998\lambda^{997} = 998\lambda^{997}\left(2\lambda^{998} + 1\right),$$

the polynomials P_A' and P_A do not have common roots. Therefore, all the roots of $P_A(\lambda)$ are distinct, thus, A has an eigenbasis.

8 First columns of the matrices $A_1, \ldots, A_{n+1}$ are linearly dependent. Therefore, for some numbers $a_1, \ldots, a_{n+1}$, not all of which are zeros, the first column of the matrix $a_1 A_1 + \ldots + a_{n+1} A_{n+1}$ is zero, so the latter matrix is singular.

9 Call the matrix $A = (a_{ij})_{i,j=1}^n$ good if it holds $a_{kk} = a_{kl} = -a_{lk} = -a_{ll} \neq 0$ for some $1 \leq k \neq l \leq n$, and all other entries are zeros. It is easy to check that the square of each good matrix is the zero matrix. Clearly, every matrix with zero trace can be written as a sum of several good matrices and a matrix with zero diagonal. It remains to represent the latter matrix as a sum of matrices with exactly one nonzero entry, which is off-diagonal.

1997

1 Each decomposition $n = a_1 + a_2 + \ldots + a_m$ corresponds to collection of partial sums:

$$b_0 = 0,\ b_1 = a_1,\ b_2 = a_1 + a_2,\ \ldots,\ b_{m-1} = a_1 + \ldots + a_{m-1},\ b_m = n.$$

Each summand k corresponds to adjacent numbers of the collection which differ by k. Fix the numbers c and $c + k$, and find the number of collections in which the numbers c and $c + k$ are adjacent. It is evident that c can take values from 0 to $n - k$. If $c = 0$ then the collection contains the numbers $0, k, n$ and some subset of the set $\{k + 1 \le i \le n - 1\}$, which can be selected in 2^{n-k-1} ways. If $c = n - k$ then the collection contains the numbers $0, n - k, n$ and some subset of the set $\{1 \le i \le n - k - 1\}$, which can be selected in 2^{n-k-1} ways. Finally for $0 < c < n - k$ the collection contains $0, c, c + k, n$ and some subsets of both sets

$$\{1 \le i \le c - 1\} \quad \text{and} \quad \{c + k + 1 \le i \le n - 1\}.$$

For each of $(n-k-1)$ values of c, we have 2^{n-k-2} ways to select a couple of subsets. Thus, a total number of summands k in the decompositions is

$$2 \cdot 2^{n-k-1} + (n - k - 1) \cdot 2^{n-k-2} = (n - k + 3) \cdot 2^{n-k-2}.$$

2 Let $K \subset \mathbb{Q}(x)$ be a field of all even rational functions over $\mathbb{Q}$, and $F \subset \mathbb{Q}(x)$ be a field of all rational functions $f(x)$ over $\mathbb{Q}$ such that $f(x + 1)$ is an even function, i.e. $f(x + 1) = f(-x + 1)$. It is evident that K and F are fields different from $\mathbb{Q}(x)$, because $x \notin K$ and $x \notin F$. Show that $[\mathbb{Q}(x) : K] = 2$. Indeed, each function $f \in \mathbb{Q}(x)$ can be expressed as

$$\tfrac{1}{2}(f(x) + f(-x)) + x \cdot \tfrac{1}{2x} \cdot (f(x) - f(-x)) = f_1(x) + x f_2(x),$$

V. Brayman and A. Kukush, *Undergraduate Mathematics Competitions (1995–2016)*, Problem Books in Mathematics,
DOI 10.1007/978-3-319-58673-1_25

where $f_1(x) = \frac{1}{2}(f(x) + f(-x)) \in K$ and $f_2(x) = \frac{1}{2x}(f(x) - f(-x)) \in K$. Similarly $[\mathbb{Q}(x) : F] = 2$. Consider $K \cap F$. If $f \in K \cap F$ then for each x it holds

$$f(x+2) = f(1 + (x+1)) = f(1 - (x+1)) = f(-x) = f(x),$$

i.e., the function f has period 2. But the only periodic rational functions are constants, thus, $K \cap F = \mathbb{Q}$. It remains to notice that $[\mathbb{Q}(x) : \mathbb{Q}] = \infty$, because $\mathbb{Q}(x)$ contains the functions $1, x, x^2, \ldots, x^n, \ldots$, which are linearly independent over $\mathbb{Q}$.

3 Consider $J_A = S^{-1}AS$, the Jordan normal form of the matrix A. Notice that matrices A and B commute if and only if the matrices $S^{-1}AS$ and $S^{-1}BS$ commute, for each polynomial $P(x)$ it holds $P(J_A) = S^{-1}P(A)S$, and the rank of a matrix does not depend on the choice of a basis. Therefore, it suffices to prove the statement for the matrix J_A.

Let the rank of $J_A - aI$ be less than $n - 1$. Then J_A consists of several Jordan cells, and each polynomial of J_A is a direct sum of polynomials of the Jordan cells. The matrix $B = (b_{ij})$, where $b_{1n} = 1$ and all other entries are zeros, commutes with J_A (in fact, $J_A B = B J_A = a \cdot B$), but it is not a polynomial of J_A.

Let the rank of $J_A - aI$ be equal to $n - 1$. Then J_A is a single Jordan cell of size n and $J_0 = J_A - aI$ is the nilpotent Jordan cell of size n. If $J_A B = B J_A$ then $(J_A - aI)B = B(J_A - aI)$, i.e., matrices which commute with J_A also commute with J_0. For $B = (b_{ij})$ the condition $J_0 B = B J_0$ implies that $b_{i,j-1} = b_{i+1,j}$, $1 \le i, j \le n$, where $b_{i0} = b_{n+1,j} = 0$. Hence $b_{ij} = 0$ for $i > j$, and $b_{ij} = \tilde{b}_{j-i}$, $j \ge i$. For each polynomial $P(x)$ we have $P(J_A) = (p_{ij})$, where

$$p_{ij} = \begin{cases} 0, & \text{if } i > j, \\ \frac{P^{(j-i)}(a)}{(j-i)!}, & \text{if } j \ge i. \end{cases}$$

Thus, if matrix B commutes with J_A then $B = P(J_A)$, where $P(x) = \sum_{k=0}^{n} \tilde{b}_k (x-a)^k$. Hence J_A commutes only with polynomials of J_A.

4 Numbers $x = 0$, $x = 1$, and $x = 3$ satisfy the equation. Show that there are no other solutions. Indeed, suppose that for the function

$$f(x) = 2^x - \frac{2}{3}x^2 - \frac{1}{3}x - 1,$$

there exist points $x_1 < x_2 < x_3 < x_4$ such that $f(x_i) = 0$, $i = 1, 2, 3, 4$. Then by Rolle's theorem there exist points $x_1 < y_1 < x_2 < y_2 < x_3 < y_3 < x_4$ such that $f'(y_i) = 0$, $i = 1, 2, 3$, points $y_1 < z_1 < y_2 < z_2 < y_3$ such that $f''(z_1) = f''(z_2) = 0$, and a point $z_1 < t < z_2$, such that $f'''(t) = 0$. But $f'''(x) = (\ln 2)^3 \cdot 2^x > 0$ for all $x \in \mathbb{R}$, a contradiction.
Answer: 0, 1, 3.

5 Fix an arbitrary $\varepsilon > 0$. Since the function $y = e^t$ is convex, for $t \in [0, \varepsilon]$ its graph lies above the tangent $y = 1 + t$ but under the secant $y = 1 + a_\varepsilon t$, where $a_\varepsilon = \frac{1}{\varepsilon}(e^\varepsilon - 1)$. Hence for $\frac{1}{n} \le \varepsilon$ it holds

$$\left(\int_0^1 e^{x^2/n} dx\right)^n \ge \left(\int_0^1 \left(1 + \frac{x^2}{n}\right) dx\right)^n = \left(1 + \frac{1}{3n}\right)^n \to e^{1/3}, \text{ as } n \to \infty,$$

and

$$\left(\int_0^1 e^{x^2/n} dx\right)^n \le \left(\int_0^1 \left(1 + \frac{a_\varepsilon x^2}{n}\right) dx\right)^n = \left(1 + \frac{a_\varepsilon}{3n}\right)^n \to e^{a_\varepsilon/3}, \text{ as } n \to \infty.$$

Since $a_\varepsilon = \frac{1}{\varepsilon}(e^\varepsilon - 1) \to 1$, as $\varepsilon \to 0+$, we get $\lim\limits_{n\to\infty} \left(\int_0^1 e^{x^2/n} dx\right)^n = e^{1/3}$.
Answer: $e^{1/3}$.

6 The matrix $I + aa^{\mathrm{T}}$ is positive definite, therefore, there exists the inverse matrix $\left(I + aa^{\mathrm{T}}\right)^{-1}$. Notice that the expressions $1 - a^{\mathrm{T}}\left(I + aa^{\mathrm{T}}\right)^{-1} a$ and $1 + a^{\mathrm{T}} a$ take real values. We have

$$\left(1 + a^{\mathrm{T}} a\right)\left(1 - a^{\mathrm{T}}\left(I + aa^{\mathrm{T}}\right)^{-1} a\right) =$$
$$= 1 + a^{\mathrm{T}} a - a^{\mathrm{T}}\left(I + aa^{\mathrm{T}}\right)^{-1} a - a^{\mathrm{T}}\left(aa^{\mathrm{T}} + I - I\right)\left(I + aa^{\mathrm{T}}\right)^{-1} a =$$
$$= 1 + a^{\mathrm{T}} a - a^{\mathrm{T}}\left(I + aa^{\mathrm{T}}\right)^{-1} a - a^{\mathrm{T}}\left(I + aa^{\mathrm{T}}\right)\left(I + aa^{\mathrm{T}}\right)^{-1} a + a^{\mathrm{T}}\left(I + aa^{\mathrm{T}}\right)^{-1} a = 1,$$

thus, $\left(1 - a^{\mathrm{T}}\left(I + aa^{\mathrm{T}}\right)^{-1} a\right)^{-1} = 1 + a^{\mathrm{T}} a.$

7 Taking into account the periodicity and signs of the functions $\sin x$ and $\cos x$, it suffices to consider $f(x)$ for $x \in \left[0, \frac{\pi}{2}\right]$ only. A necessary condition for a maximum is

$$f'(x) = e^{\sin x}\cos x - e^{\cos x}\sin x = 0,$$

hence $x \in \left(0, \frac{\pi}{2}\right)$ and $\frac{e^{\sin x}}{\sin x} = \frac{e^{\cos x}}{\cos x}$. The function $g(t) = \frac{e^t}{t}$ decreases on $(0, 1)$, because $g'(t) = \frac{e^t(t-1)}{t^2} < 1$. Therefore, the latter equality is equivalent to $\sin x = \cos x$, so $x = \frac{\pi}{4}$. Thus, $x = \frac{\pi}{4}$ is a unique extreme point in $\left[0, \frac{\pi}{2}\right]$ and

$$\max_{x\in\mathbb{R}} f(x) = \max_{x\in\left[0,\frac{\pi}{2}\right]} f(x) = f\left(\tfrac{\pi}{4}\right) = 2e^{\frac{1}{\sqrt{2}}}.$$

Answer: $2e^{\frac{1}{\sqrt{2}}}$.

8 Denote $I_k = \left[k\pi - \frac{\pi}{2}, k\pi + \frac{\pi}{2}\right]$, $k \ge 1$. Estimate the integral $\displaystyle\int_{I_k} \frac{f(x)}{|\sin x|^{1-\frac{1}{x}}} dx$.
The function $|\sin x|$ is concave on both segments $\left[k\pi - \frac{\pi}{2}, k\pi\right]$ and $\left[k\pi, k\pi + \frac{\pi}{2}\right]$,

hence $|\sin x| \geq \frac{2}{\pi}|x - k\pi|$, $x \in I_k$. For $x \in I_k$, it also holds $1 - \frac{1}{x} < 1 - \frac{1}{(k+1)\pi}$ and $f(x) \leq f\left(k\pi - \frac{\pi}{2}\right)$. Thus,

$$|\sin x|^{1-\frac{1}{x}} \geq |\sin x|^{1-\frac{1}{(k+1)\pi}} \geq \left(\frac{2}{\pi}|x - k\pi|\right)^{1-\frac{1}{(k+1)\pi}} \geq \frac{2}{\pi}|x - k\pi|^{1-\frac{1}{(k+1)\pi}}, \; x \in I_k.$$

Therefore,

$$\begin{aligned}
&\int_{I_k} \frac{f(x)}{|\sin x|^{1-\frac{1}{x}}} dx \leq \frac{\pi}{2} f\left(k\pi - \frac{\pi}{2}\right) \cdot \int_{I_k} \frac{dx}{|x - k\pi|^{1-\frac{1}{(k+1)\pi}}} = \\
&= \frac{\pi}{2} f\left(k\pi - \frac{\pi}{2}\right) \int_{-\frac{\pi}{2}}^{\frac{\pi}{2}} |x|^{\frac{1}{(k+1)\pi}-1} dx = \\
&= \pi f\left(k\pi - \frac{\pi}{2}\right) \cdot (k+1)\pi \cdot \left(\frac{\pi}{2}\right)^{\frac{1}{(k+1)\pi}} \leq \frac{\pi^3}{2} f\left(k\pi - \frac{\pi}{2}\right)(k+1).
\end{aligned}$$

Thus,

$$\int_1^\infty \frac{f(x)dx}{|\sin x|^{1-\frac{1}{x}}} \leq \int_1^{\frac{\pi}{2}} \frac{f(x)dx}{|\sin x|^{1-\frac{1}{x}}} + \frac{\pi^3}{2} \sum_{k=1}^{\infty} f\left(k\pi - \frac{\pi}{2}\right)(k+1).$$

It remains to prove convergence of the series $\sum_{k=1}^{\infty} f\left(k\pi - \frac{\pi}{2}\right)(k+1)$. For every $k \geq 4$ it holds

$$f\left(k\pi - \frac{\pi}{2}\right)(k+1)\pi \leq 2f\left(k\pi - \frac{\pi}{2}\right)\left(k\pi - \frac{3\pi}{2}\right) \leq \int_{I_{k-1}} x f(x)\, dx.$$

Therefore, convergence of the integral $\int_1^\infty x f(x)\, dx$ implies convergence of the series.

9 Rewrite the expression as an integral sum:

$$\sum_{j=1}^{n^2} \frac{n}{n^2 + j^2} = \sum_{j=1}^{n^2} \frac{1}{n} \cdot \frac{1}{1 + \left(\frac{j}{n}\right)^2}.$$

Since for $x \in \left[\frac{j}{n}, \frac{j+1}{n}\right]$ the inequality

$$\frac{1}{1 + \left(\frac{j+1}{n}\right)^2} \leq \frac{1}{1 + x^2} \leq \frac{1}{1 + \left(\frac{j}{n}\right)^2}$$

holds, we have $\int_{\frac{1}{n}}^{n+\frac{1}{n}} \frac{dx}{1+x^2} \le \sum_{j=1}^{n^2} \frac{n}{n^2+j^2} \le \int_0^n \frac{dx}{1+x^2}$. Hence

$$\lim_{n\to\infty} \sum_{j=1}^{n^2} \frac{n}{n^2+j^2} = \int_0^\infty \frac{dx}{1+x^2} = \frac{\pi}{2}.$$

Answer: $\frac{\pi}{2}$.

10 Assume that $|f'(x_0)| > 2$ for some point $x_0 \in I = (a, b)$. Without loss of generality $f'(x_0) \ge 2 + \varepsilon$, where $0 < \varepsilon < 1$. Then for every $x \in (a, b)$ it holds

$$f'(x) = f'(x_0) + \int_{x_0}^{x} f''(t)\, dt \ge 2 + \varepsilon - |x - x_0|.$$

Select points $a < a_1 < x_0$ and $x_0 < b_1 < b$ such that $2 - \varepsilon < b_1 - a_1 \le 2$. Then

$$\begin{aligned} f(b_1) - f(a_1) = &\int_{a_1}^{b_1} f'(t)\, dt \ge \int_{a_1}^{b_1} (2 + \varepsilon - |t - x_0|)\, dt = \\ =&(2+\varepsilon)(b_1 - a_1) - \frac{(a_1 - x_0)^2}{2} - \frac{(b_1 - x_0)^2}{2}. \end{aligned}$$

Find the minimum of the latter expression for $x_0 \in (a_1, b_1)$ and obtain

$$\begin{aligned} f(b_1) - f(a_1) \ge &(2+\varepsilon)(b_1 - a_1) - \frac{(b_1 - a_1)^2}{2} > \\ > &(2 + \varepsilon - 1)(2 - \varepsilon) = 2 + \varepsilon - \varepsilon^2 > 2, \end{aligned}$$

because $\varepsilon < 1$. On the other hand, $|f(b_1)| \le 1$ and $|f(a_1)| \le 1$, so $f(b_1) - f(a_1) \le 2$, and we get a contradiction. Thus, $|f'(x)| \le 2$ on I.

11 Let $\deg P = n$ and $\deg Q = k$, $n \ge k$. Then the degree of the polynomial $P - Q$ does not exceed n. Find the number of roots of $P - Q$. Let $x_1, \dots, x_m$ be the roots of the polynomial P of multiplicities $s_1, \dots, s_m$, and $y_1, \dots, y_l$ be the roots of the polynomial $P + 1$ of multiplicities $t_1, \dots, t_l$, respectively, where $s_1 + \ldots + s_m = t_1 + \ldots + t_l = n$. Then the polynomial $P - Q$ has at least $m + l$ distinct roots, namely $x_1, \dots, x_m, y_1, \dots, y_l$. On the other hand, $x_1, \dots, x_m$ and $y_1, \dots, y_l$ are the roots of P' of multiplicities $s_1 - 1, \dots, s_m - 1$ and $t_1 - 1, \dots, t_l - 1$, respectively, thus,

$$s_1 - 1 + \ldots + s_m - 1 + t_1 - 1 + \ldots + t_l - 1 = n - m + n - l \le n - 1.$$

Hence $m + l \ge n + 1$, so the polynomial $P - Q$ of degree at most n has at least $n + 1$ roots. Thus, $P - Q \equiv 0$.

1998

1 Denote by $\mathbb{N}_n$ the set $\{1, 2, \ldots, n\}$ and by f_n the number of minimal selfish subsets of $\mathbb{N}_n$. Then the number of minimal selfish subsets of $\mathbb{N}_n$, which do not contain n, equals f_{n-1}. On the other hand, each minimal selfish subset of $\mathbb{N}_n$, which contains n, allows to obtain a minimal selfish subset of $\mathbb{N}_{n-2}$ by subtracting 1 from each elements and then removing the element $n-1$. It is possible, because minimal selfish subset except $\{1\}$ cannot contain 1. The inverse procedure maps each minimal selfish subset of $\mathbb{N}_{n-2}$ to a minimal selfish subset of $\mathbb{N}_n$, which contains n. Hence, $f_n = f_{n-1} + f_{n-2}$. Since $f_1 = f_2 = 1$, we get $f_n = F_n$, where F_n is the nth Fibonacci number.
Answer: the number of subsets in question equals to the nth Fibonacci number.

2 Let binary expansion of α be $0.a_1a_2\ldots a_n\ldots$ We will toss the coin till the first head appears. If it appears first at the nth try, then we regard that the first player wins if $a_n = 1$ and loses if $a_n = 0$. The probability that the game ends exactly after n tosses equals $\frac{1}{2^n}$. Therefore, the game ends after finite number of moves with probability $\sum\limits_{n=1}^{\infty} \frac{1}{2^n} = 1$, and moreover the first player wins with probability

$$\sum_{n=1}^{\infty} \frac{a_n}{2^n} = 0.a_1a_2\ldots a_n\ldots = \alpha.$$

Answer: yes.

3 Place the triangle on the Cartesian plane so that its vertices have coordinates $C(0,0)$, $A(4,0)$, and $B(0,3)$. Also consider points $D\left(\frac{27}{13}, 0\right)$ and $E\left(\frac{20}{13}, \frac{24}{13}\right)$. It is easy to verify that D lies on AC, E lies on AB, and $BE = DE = AD = \frac{25}{13}$ (see Fig. 1). Since some part of the dissection contains at least two of the points A, B, C, D, E, and all the distances between these points are not less than $\frac{25}{13}$, the diameter of the dissection is not less than $\frac{25}{13}$.

V. Brayman and A. Kukush, *Undergraduate Mathematics Competitions (1995–2016)*, Problem Books in Mathematics,
DOI 10.1007/978-3-319-58673-1_26

Give an example of dissection with diameter $\frac{25}{13}$. Consider points $F\left(\frac{32}{13}, \frac{15}{13}\right)$, $G(1, 0)$, $H\left(\frac{7}{13}, \frac{15}{13}\right)$, $I\left(0, \frac{20}{13}\right)$, and divide the triangle ABC into parts AFD, $EFDGH$, $BEHI$, and $CGHI$. Then $BEDH$ and $ADHF$ are rhombuses with the side $\frac{25}{13}$. It is easy to verify that all the distances between other vertices of polygons of the dissection not exceed $\frac{25}{13}$, so the diameter of the dissection equals $\frac{25}{13}$.
Answer: $\frac{25}{13}$.

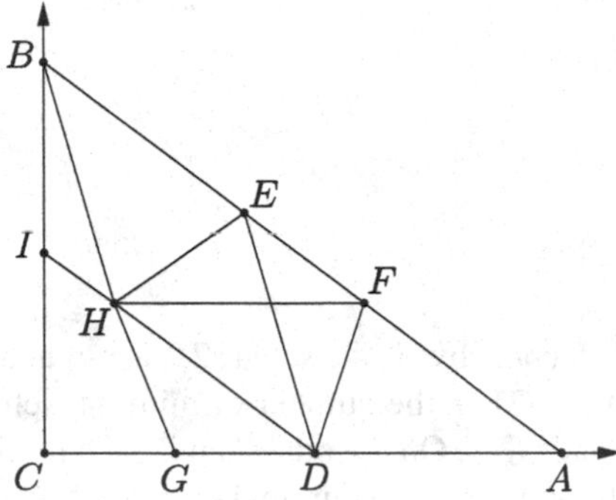

Fig. 1 The dissection of the triangle

4 One can take $B = \frac{1}{1-q} \cdot A^{-1}$.

5 First show that $A^2 = \pm I$ for all $A \in M$. Indeed, for arbitrary $B \in M$ it holds $BA = kAB$, where $k^2 = 1$, thus, $BA^2 = kABA = k^2A^2B = A^2B$. Therefore, A^2 commutes with all $B \in M$ and $-A^2$ has the same property. By condition (ii) either A^2 or $-A^2$ belongs to M, so either $A^2 = \pm I$ or we get a contradiction with condition (iv).

If M contains more than n^2 matrices, then the matrices are linearly dependent, i.e., $\sum\limits_{A \in M} \lambda_A A = O$ for some real numbers λ_A not all of which are zeros. Consider a nontrivial collection λ_A, $A \in M$, which contains the least possible number of nonzero coefficients. Let $\lambda_{A_0} \neq 0$. Then $\sum\limits_{A \in M} \lambda_A A A_0 = O$, $A_0^2 = \pm I$, and for each matrix $A \in M$, either $AA_0 \in M$ or $-AA_0 \in M$. Thus, we found a new linear combination with the least possible number of nonzero coefficients, moreover now $\lambda_I \neq 0$. There exists at least one matrix $A_1 \in M$, for which $A_1 \neq I$ and $\lambda_{A_1} \neq 0$. Select a matrix $B \in M$ from condition (iv) such that $A_1 B = -BA_1$. We have $\sum\limits_{A \in M} \lambda_A (AB + BA) = O$. For each $A \in M$ it holds either $AB + BA = 0$, or $\frac{1}{2}(AB + BA) \in M$, or $-\frac{1}{2}(AB + BA) \in M$. In particular if $A = A_1$ then $AB + BA = 0$, and if $A = I$ then $\frac{1}{2}(AB + BA) = B \in M$. Therefore, the left-hand side of the equality $\sum\limits_{A \in M} \lambda_A (AB + BA) = O$ can be rewritten as a nontrivial linear combination of matrices from M with less number of nonzero coefficients. We get a contradiction.

6 In the field with two elements, 2=0 and $a_n^2 = a_n$. Therefore,

$$\alpha^2 = 1 + \sum_{n\geq 1} a_n^2 x^{2n} = 1 + \sum_{n\geq 1} a_n x^{2n}, \quad \alpha^4 = 1 + \sum_{n\geq 1} a_n x^{4n}.$$

Find $\alpha^4 + x\alpha^2 + \alpha$. Consider separately the coefficients at x^{4n}, x^{4n+2}, and x^{2n+1}. The coefficient at x^{4n} equals $a_n + a_{4n}$ if $n \geq 1$, and equals $1 + 1 = 0$ if $n = 0$. Binary expansion of the number $4n$ is obtained by appending 00 to the expansion of n. Therefore, $a_{4n} = a_n$, so $a_{4n} + a_n = 0$. The coefficient at x^{4n+2} equals $a_{4n+2} = 0$, because binary expansion of $4n+2$ ends with 10. At last, the coefficient at x^{2n+1} equals $a_{2n+1} + a_n$. The expansion of $2n+1$ is obtained from the expansion of n by appending 1, hence $a_{2n+1} = a_n$, so $a_{2n+1} + a_n = 0$.

Thus, $\alpha^4 + x\alpha^2 + \alpha = 0$. Since $\alpha \neq 0$, we have $\alpha^3 + x\alpha + 1 = 0$.

7 Introduce a function $h(x) = (f(x))^2 + \left(f'(x)\right)^2$. It holds

$$h'(x) = 2f(x)f'(x) + 2f'(x)f''(x) = -xg(x)\left(f'(x)\right)^2.$$

Hence, the function $h(x)$ is increasing for $x < 0$ and decreasing for $x > 0$. Therefore, $(f(x))^2 \leq h(x) \leq h(0)$ for all $x \in \mathbb{R}$.

8 Construct a function $F \in C^{(1)}(\mathbb{R})$ such that

$$\forall t \in \mathbb{R} \quad F(t+1) = F(t) + \arctan t.$$

Since $\int_0^1 f(x+t)\,dx = \int_t^{t+1} f(x)\,dx$, the function $f(t) = F'(t)$ will satisfy the condition of the problem. First define F on the interval $[0, 1)$, and then define F on the intervals $[1, 2), [2, 3), \ldots$ and $[-1, 0), [-2, -1), \ldots$ in such a way that the condition

$$F(t+1) = F(t) + \arctan t$$

holds. It suffices to ensure that $F \in C^{(1)}(\mathbb{R})$. Continuity of F and F' at point $t = 1$ implies that $F(1) = F(1-)$, $F'(1-) = F'(1+)$, i.e.,

$$F(1-) = F(0) + \arctan 0 = F(0),$$

$$F'(1-) = F'(0+) + (\arctan t)'\big|_{t=0} = F'(0+) + 1.$$

These conditions hold if, e.g., $F(t) = \frac{1}{2}(t^2 - t)$ for $t \in [0, 1)$. Then $F \in C^{(1)}((0, 2))$ and by the construction F and F' are continuous at all integer points. Therefore, $F \in C^{(1)}(\mathbb{R})$.

9 Since $\varphi(g)\varphi(e)\varphi(g^{-1}) = \varphi(e)\varphi(g)\varphi(g^{-1})$, elements $\varphi(g)$ and $\varphi(e)$ commute for all $g \in G$. Hence, $\varphi(g)$ and $\varphi(e)^{-1}$ commute. Also we have

$$\varphi(x)\varphi(y)\varphi\left(y^{-1}x^{-1}\right) = \varphi(e)\varphi(xy)\varphi\left(y^{-1}x^{-1}\right),$$

whence $\varphi(x)\varphi(y) = \varphi(e)\varphi(xy)$. Therefore,

$$\varphi(e)^{-1}\varphi(x)\varphi(e)^{-1}\varphi(y) = \varphi(e)^{-1}\varphi(e)^{-1}\varphi(x)\varphi(y) = \varphi(e)^{-1}\varphi(xy),$$

i.e., for $\psi(x) = \varphi(e)^{-1}\varphi(x)$ it holds $\psi(x)\psi(y) = \psi(xy)$, $x, y \in G$.

10 Show that the limit in question does not change if we replace $\{x_n\}$ with a new sequence $\{y_n\}$, defined as

$$y_1 = 1, \quad y_{n+1} = \frac{1}{2+y_n} + \xi_n, \; n \geq 1,$$

where $\xi_n \in [0, 1]$, $n \geq 1$, are arbitrary numbers such that $\{\sqrt{n}\} - \xi_n \to 0$, as $n \to \infty$. Indeed,

$$\begin{aligned} |y_{n+1} - x_{n+1}| &\leq \left|\frac{1}{2+x_n} - \frac{1}{2+y_n}\right| + \left|\{\sqrt{n}\} - \xi_n\right| \leq \\ &\leq \frac{1}{4}\,|y_n - x_n| + \left|\{\sqrt{n}\} - \xi_n\right|, \; n \geq 1, \end{aligned}$$

thus,

$$|y_n - x_n| \leq \sum_{k=1}^{n-1} \frac{|\{\sqrt{n}\} - \xi_n|}{4^{n-k-1}} \to 0, \text{ as } n \to \infty.$$

It is easy to verify that $x_n, y_n \in [0, 3/2]$, $n \geq 1$. Therefore,

$$|y_n^2 - x_n^2| \leq 3|y_n - x_n| \to 0, \text{ as } n \to \infty,$$

and by Toeplitz's theorem $\lim\limits_{N\to\infty} \frac{1}{N} \sum\limits_{k=1}^{N} |x_k^2 - y_k^2| = 0.$

For $M \geq 1$, $M^2 \leq N \leq (M+1)^2 - 1$ and $n = N^2 + a$, $0 \leq a \leq 2N$, set $\xi_n = \frac{1}{M}\left[\frac{aM}{2N}\right]$. Then $\{\xi_n, n \geq 1\} \subset [0, 1]$. Since

$$N + \tfrac{a-1}{2N} < \sqrt{N^2 + a} < N + \tfrac{a}{2N},$$

we have $\left|\frac{a}{2N} - \left\{\sqrt{N^2+a}\right\}\right| < \frac{1}{2N}$. Also we have $\left|\frac{a}{2N} - \frac{1}{M}\left[\frac{aM}{2N}\right]\right| < \frac{1}{M}$, whence

$$\left|\{\sqrt{n}\} - \xi_n\right| \leq \frac{1}{2N} + \frac{1}{M} \to 0, \text{ as } n \to \infty,$$

because $N, M \to \infty$, as $n \to \infty$.

Notice that the finite sequence $\xi_{N^2}, \xi_{N^2+1}, \dots, \xi_{N^2+2N}$ consists of numbers $0, \frac{1}{M}, \dots, \frac{M-1}{M}$, each of which repeats $\left[\frac{2N+1}{M}\right]$ times or $\left[\frac{2N+1}{M}\right]+1$ times in a row, i.e., the sequence $\{\xi_n,\ n \geq 1\}$ consists of constant parts, lengths of which tend to infinity.

For each fixed $\xi \in [0,1]$, an equation $x(\xi) = \frac{1}{2+x(\xi)} + \xi$ has a unique solution in $[0, 3/2]$, namely $x(\xi) = \frac{\xi-2+\sqrt{\xi^2+4\xi+8}}{2}$. For arbitrary $z_1 \in [0, 3/2]$, define a sequence $\{z_n\}$ as

$$z_{n+1} = \frac{1}{2+z_n} + \xi,\ n \geq 1.$$

Then

$$|z_n - x(\xi)| = \left|\frac{1}{2+z_{n-1}} - \frac{1}{2+x(\xi)}\right| \leq$$
$$\leq \frac{1}{4}|z_{n-1} - x(\xi)| \leq \dots \leq \frac{3}{2\cdot 4^{n-1}},\ n \geq 2.$$

Thus, $|z_n^2 - x^2(\xi)| \leq \frac{9}{2\cdot 4^{n-1}}$, $n \geq 1$, and $\sum\limits_{k=1}^{n} |z_k^2 - x^2(\xi)| \leq \sum\limits_{k=1}^{\infty} \frac{9}{2\cdot 4^{k-1}} = 6$.

Fix $M \geq 1$ and $M^2 \leq N \leq (M+1)^2 - 1$. Consider an average $\frac{1}{2N+1} \sum\limits_{k=N^2+1}^{(N+1)^2} y_k^2$. The numbers $y_{N^2+1}, \dots, y_{(N+1)^2}$ form M finite sequences of length either $\left[\frac{2N+1}{M}\right]$ or $\left[\frac{2N+1}{M}\right]+1$, each of which is constructed by the same rule as z_n, where ξ takes values $0, \frac{1}{M}, \dots, \frac{M-1}{M}$. Therefore,

$$\left|\frac{1}{2N+1}\sum_{k=N^2+1}^{(N+1)^2} y_k^2 - \frac{1}{M}\sum_{j=0}^{M-1} x^2\left(\frac{j}{M}\right)\right| \leq \frac{1}{2N+1}\sum_{k=N^2+1}^{(N+1)^2} |y_k^2 - x^2(\xi_{k-1})| +$$
$$+\left|\frac{1}{2N+1}\sum_{k=N^2+1}^{(N+1)^2} x^2(\xi_{k-1}) - \frac{1}{M}\sum_{j=0}^{M-1} x^2\left(\frac{j}{M}\right)\right| \leq$$
$$\leq \frac{6M}{2N+1} + \frac{9}{4}\cdot\frac{M}{2N+1} = \frac{33M}{4(2N+1)}.$$

It holds

$$M = \left[\sqrt{N}\right] \to \infty,\ \frac{M}{2N+1} \leq \frac{\sqrt{N}}{N} \to 0,\ \frac{1}{M}\sum_{j=0}^{M-1} x^2\left(\frac{j}{M}\right) \to \int_0^1 x^2(\xi)d\xi,\ \text{as } N \to \infty.$$

Thus, $\left|\frac{1}{2N+1}\sum\limits_{k=N^2+1}^{(N+1)^2} y_k^2 - \int_0^1 x^2(\xi)d\xi\right| \to 0$, as $N \to \infty$.

By Toeplitz's theorem,

$$\left|\frac{1}{L^2}\sum_{k=1}^{L^2} y_k^2 - \int_0^1 x^2(\xi)d\xi\right| \le \frac{1}{L^2}\sum_{N=0}^{L-1}(2N+1)\times$$

$$\times\left|\frac{1}{2N+1}\sum_{k=N^2+1}^{(N+1)^2} y_k^2 - \int_0^1 x^2(\xi)d\xi\right| \to 0, \text{ as } L \to \infty.$$

At last for all integers n, $L^2 \le n < (L+1)^2$, it holds

$$\left|\frac{1}{n}\sum_{k=1}^{n} y_k^2 - \int_0^1 x^2(\xi)d\xi\right| \le \frac{L^2}{n}\left|\frac{1}{L^2}\sum_{k=1}^{L^2} y_k^2 - \int_0^1 x^2(\xi)d\xi\right| +$$

$$+\frac{n-L^2}{n}\left(\frac{9}{4}+\int_0^1 x^2(\xi)d\xi\right) \to 0, \text{ as } n \to \infty,$$

because $\frac{L^2}{n} \to 1$, $\frac{n-L^2}{n} \to 0$, as $n \to \infty$. Therefore,

$$\lim_{n\to\infty}\frac{1}{n}\sum_{k=1}^{n} x_k^2 = \lim_{n\to\infty}\frac{1}{n}\sum_{k=1}^{n} y_k^2 =$$

$$= \int_0^1 x^2(\xi)d\xi = \int_0^1 \left(\frac{\xi-2+\sqrt{\xi^2+4\xi+8}}{2}\right)^2 d\xi.$$

It remains to compute the integral

$$\int_0^1 \left(\frac{t-2+\sqrt{t^2+4t+8}}{2}\right)^2 dt = \frac{1}{4}\int_0^1 \left(t-2+\sqrt{t^2+4t+8}\right)^2 dt =$$

$$= \frac{1}{4}\int_0^1 \left(2t^2+12+2(t-2)\sqrt{t^2+4t+8}\right) dt =$$

$$= 3\frac{1}{6} + \frac{1}{2}\int_2^3 (s-4)\sqrt{s^2+4}\,ds =$$

$$= 3\frac{1}{6} + \frac{1}{2}\int_2^3 s\sqrt{s^2+4}\,ds - 2\int_2^3 \sqrt{s^2+4}\,ds =$$

$$= 3\frac{1}{6} + \frac{1}{6}\left(s^2+4\right)^{3/2}\Big|_2^3 - 2\left(2\ln\left(s+\sqrt{s^2+4}\right)+\frac{s}{2}\sqrt{s^2+4}\right)\Big|_2^3 =$$

$$= 3\frac{1}{6} - \frac{5}{6}\sqrt{13} + \frac{4}{3}\sqrt{2} - 4\ln\frac{3+\sqrt{13}}{2+2\sqrt{2}}.$$

Answer: $\lim_{n\to\infty}\frac{1}{n}\sum_{k=1}^{n} x_k^2 = 3\frac{1}{6} - \frac{5}{6}\sqrt{13} + \frac{4}{3}\sqrt{2} - 4\ln\frac{3+\sqrt{13}}{2+2\sqrt{2}}.$

11 Consider a matrix A^{T} of size $n \times n$, where $A = (a_{ij})$. Since $\operatorname{tr} A^{\mathrm{T}} > 0$, there exists λ, an eigenvalue of A^{T} such that $\operatorname{Re}\lambda > 0$. Consider the corresponding eigenvector $v = (v_i)$ and a linear combination $y = \sum\limits_{i=1}^{n} v_i x_i$. We have

$$y' = \sum_{i=1}^{n} v_i \cdot \sum_{j=1}^{n} a_{ij}x_j = \sum_{j=1}^{n}\left(\sum_{i=1}^{n} a_{ij}v_i\right)x_j = \lambda \sum_{j=1}^{n} v_j x_j = \lambda y,$$

thus, $y(t) = ke^{\lambda t}$ for some constant k. But $|e^{\lambda t}|$ does not tend to zero, as $t \to \infty$, because $\operatorname{Re}\lambda > 0$. Since $y(t) = \sum\limits_{i=1}^{n} v_i x_i(t) \to 0$, as $t \to \infty$, we have $k = 0$. Thus, $\sum\limits_{i=1}^{n} v_i x_i(t) = 0$, and the functions $x_1, x_2, \ldots, x_n$ are linearly dependent.

12 The statement of the problem is a corollary of the following more general statement: for arbitrary operators $A_1, A_2 \in \mathcal{L}(B)$

$$\sigma(A_1A_2)\backslash\{0\} = \sigma(A_2A_1)\backslash\{0\}.$$

Prove this statement.

Let $\lambda \in \mathbb{C}\backslash\{0\}$ be such a number that there exists $(A_1A_2 - \lambda I)^{-1} \in \mathcal{L}(B)$, where I is the identity operator. It holds

$$\begin{aligned}
(A_2A_1 - \lambda I)A_2 &= A_2(A_1A_2 - \lambda I),\\
(A_2A_1 - \lambda I)A_2(A_1A_2 - \lambda I)^{-1} &= A_2,\\
(A_2A_1 - \lambda I)A_2(A_1A_2 - \lambda I)^{-1}A_1 &= A_2A_1 - \lambda I + \lambda I,\\
(A_2A_1 - \lambda I)\left(A_2(A_1A_2 - \lambda I)^{-1}A_1 - I\right) &= \lambda I,\\
(A_2A_1 - \lambda I)\left(\frac{1}{\lambda}\left(A_2(A_1A_2 - \lambda I)^{-1}A_1 - I\right)\right) &= I.
\end{aligned}$$

In a similar way $\frac{1}{\lambda}\left(A_2(A_1A_2 - \lambda I)^{-1}A_1 - I\right)(A_2A_1 - \lambda I) = I$. Thus, there exists

$$(A_2A_1 - \lambda I)^{-1} = \frac{1}{\lambda}\left(A_2(A_1A_2 - \lambda I)^{-1}A_1 - E\right) \in \mathcal{L}(B).$$

Hence, for resolvent sets it holds $\rho(A_1A_2)\backslash\{0\} = \rho(A_2A_1)\backslash\{0\}$. Therefore,

$$\sigma(A_1A_2)\backslash\{0\} = \sigma(A_2A_1)\backslash\{0\}.$$

1999

2 A solution is similar to the solution of Problem 7, 1997.
Answer: $2 \cdot 2^{\frac{1}{\sqrt{2}}}$.

3 If either f or one of the derivatives changes sign then the expression

$$f(a) \cdot f'(a) \cdot f''(a) \cdot f'''(a)$$

attains zero value. Assume that each of the functions f, f', f'', f''' is either strictly positive or strictly negative. Show that $f(x) \cdot f''(x) > 0$ for all $x \in \mathbb{R}$. Indeed, let $f''(x) > 0$, $x \in \mathbb{R}$. Then f is convex, and

$$f(x+y) \geq f(x) + yf'(x), \; x, y \in \mathbb{R}.$$

For y of the same sign as $f'(x)$, with absolute value large enough, it holds $f(x+y) > f(x) + yf'(x) > 0$. Hence $f(x) > 0$ for all $x \in \mathbb{R}$. If $f''(x) < 0$ then in a similar way we induce that the function f is negative, so $f(x) \cdot f''(x) > 0$ as well. Similarly $f'(x) \cdot f'''(x) > 0$, $x \in \mathbb{R}$. Thus, $f(x) \cdot f'(x) \cdot f''(x) \cdot f'''(x) > 0$.

4 Denote by Λ the sum of corresponding eigenvalues. Consider one of the eigenvectors with some eigenvalue λ. Other n eigenvectors are linearly independent, so they form a basis, and the trace of the matrix in this basis equals $\Lambda - \lambda$. But the trace of a matrix does not depend on the choice of a basis. Therefore, all the eigenvalues are equal and the transformation has a form λI.

5 For $m = 999$, it holds $N = \frac{1}{9}(10^{2m} - 1)$, and $\sqrt{N} = \frac{10^m}{3}\sqrt{1 - 10^{-2m}}$. It is easy to verify that

$$1 - \tfrac{2}{3} \cdot 10^{-2m} < \sqrt{1 - 10^{-2m}} < 1 - \tfrac{1}{2} \cdot 10^{-2m}.$$

V. Brayman and A. Kukush, *Undergraduate Mathematics Competitions (1995–2016)*, Problem Books in Mathematics,
DOI 10.1007/978-3-319-58673-1_27

Therefore,

$$3\ldots3.3\ldots311\ldots < \sqrt{N} < 3\ldots3.3\ldots316\ldots$$

(there are 999 threes before and 999 threes after decimal point). Thus, the thousandth digit after decimal point equals 1.
Answer: 1.

6 If $-1 \in S$ then $(-1)(-1) = 1 \in S$, and for $r = 1$ we get a contradiction. Thus, $1 \in S$, $1+1 = 2 \in S$, $1+2 = 3 \in S$, and by induction on $n \geq 1$ we get that $n \in S$ for all positive integers n. Therefore, S contains all positive integers. It is evident that S does not contain zero, otherwise for $r = 0$ all three statements are true. Suppose that S contains some negative rational number $-\frac{k}{m}$, where $k, m \in \mathbb{N}$. Again by induction on $n \geq 1$ we obtain that the set S contains all the numbers of the form $-\frac{kn}{m}$, $n \in \mathbb{N}$. Then simultaneously $-k \in S$ and $k \in S$, a contradiction. Therefore, $S \subset (0, +\infty)$, so $S = (0, +\infty) \cap \mathbb{Q}$, otherwise there exists such r that neither of three statements from the condition of the problem is true.

8 Show that the set S contains the circle which bounds the disc D. Since each interior point of the disc is the middle of the chord orthogonal to the diameter passing through the point, the statement of the problem will follow.

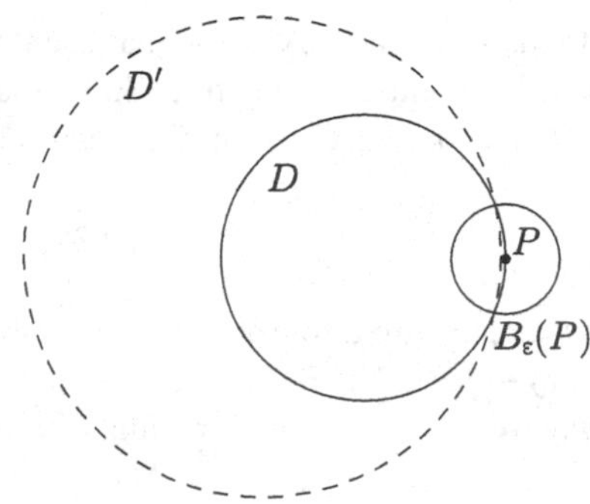

Fig. 1 A disc D' such that $D \backslash B_\varepsilon(P) \subset D'$ but $D \not\subset D'$

Take an arbitrary point P of the circle, and show that the point P belongs to S. If it is not true then some disc $B_\varepsilon(P)$ with the center P does not intersect S, because the set $\mathbb{R}^2 \backslash S$ is open. Then one can find a disc D' such that $S \subset (D \backslash B_\varepsilon(P)) \subset D'$ but $D \not\subset D'$, in particular $P \notin D'$ (see Fig. 1). We get a contradiction.

9 By the polar decomposition theorem there exist self-adjoint positive semidefinite matrices $\{P_n, n \geq 1\}$ and orthogonal matrices $\{U_n, n \geq 1\}$ such that $S_n = P_n U_n$, $n \geq 1$. We have

$$S_n S_n^{\mathrm{T}} = P_n U_n U_n^{\mathrm{T}} P_n^{\mathrm{T}} = P_n^2 \to I, \text{ as } n \to \infty.$$

Therefore, all the eigenvalues of P_n tend to 1, and $P_n \to I$, as $n \to \infty$. Hence $S_n - U_n = (P_n - I)U_n \to O$, as $n \to \infty$.

10 Denote by F_ξ and F_η the cumulative distribution functions of ξ and η. Assume that for all $a \in \mathbb{R}$ it holds $\mathsf{P}(\xi = a) \cdot \mathsf{P}(\eta = a) = 0$, that is F_ξ and F_η satisfy the equality $(F_\xi(a) - F_\xi(a-))(F_\eta(a) - F_\eta(a-)) = 0$. Then the independence of ξ and η implies that

$$\mathsf{P}(\xi = \eta) = \int_{-\infty}^{\infty} (F_\xi(a) - F_\xi(a-))dF_\eta(a) = 0,$$

a contradiction with the condition $\mathsf{P}(\xi = \eta) > 0$.

11 Let $\{e_i,\ i \geq 2\}$ be an orthobasis in H and $e_1 = \sum_{i=2}^{\infty} \frac{e_i}{i}$. Then e_1 is well-defined because for $n > m \geq 2$ it holds

$$\left\| \sum_{i=2}^{n} \frac{e_i}{i} - \sum_{i=2}^{m} \frac{e_i}{i} \right\|_H^2 = \sum_{i=m+1}^{n} \frac{1}{i^2} \to 0, \text{ as } m, n \to \infty,$$

and therefore, the series converges in H. It is evident that every nontrivial linear combination of a finite number of elements $e_{i_1}, e_{i_2}, \ldots, e_{i_k}$ does not equal zero, i.e., the elements of the set $M = \{e_i,\ i \geq 1\}$ are linearly independent. It is clear that H is the closed linear hull of the set $\{e_i,\ i \geq 2\}$. Thus, it suffices to show that e_i belongs to the closed linear hull of the set $M \backslash \{e_i\}$, for each $i \geq 2$. It is true, because for $n > i$ we have

$$\left\| e_i - i \left(e_1 - \sum_{\substack{j=2 \\ j \neq i}}^{n} \frac{e_j}{j} \right) \right\|_H^2 = i^2 \sum_{j=n+1}^{\infty} \frac{1}{j^2} \to 0, \text{ as } n \to \infty.$$

2000

1 If the series $\sum_{n=1}^{\infty} a_n$ converges to S then $b_n \le Sn$. Hence $\frac{1}{b_n} \ge \frac{1}{Sn}$, and the series $\sum_{n=1}^{\infty} \frac{1}{b_n}$ diverges.

2 Consider $M'_\alpha = [0, \alpha] \cap \mathbb{Q}$, $\alpha \in \mathscr{A} = [0, 1]$. Take an arbitrary bijection of $\mathbb{Q}$ onto $\mathbb{N}$. For each α let M_α be the image of M'_α under this bijection.
Answer: Yes, it is possible.

3 For $y = 0$, we have that $f(1) = \frac{f(x)+f(0)}{f(x)-f(0)}$ does not depend on x. Therefore, $f(0) = 0$ and $f(1) = 1$. For $x = ty$, $t > 1$, we have

$$f\left(\tfrac{t+1}{t-1}\right) = \frac{f(ty) + f(y)}{f(ty) - f(y)},$$

whence

$$\frac{f(ty)}{f(y)} = \frac{f\left(\frac{t+1}{t-1}\right) + 1}{f\left(\frac{t+1}{t-1}\right) - 1}$$

does not depend on y. Therefore,

$$f(ty) = f(y) \cdot \frac{f(t)}{f(1)} = f(y)f(t).$$

By induction on n we get $f(x^n) = f(x)^n$, $n \in \mathbb{N}$. Since $f(x^n)f(x^{-n}) = f(1) = 1$, we have $f(x^n) = f(x)^n$, $n \in \mathbb{Z}$. Let $f(2) = 2^a$, where $a > 0$ because $f(2) > f(1) = 1$. Then $f(2^{1/m})^m = f(2) = 2^a$, thus, $f(2^{1/m}) = 2^{a/m}$, so $f(2^{n/m}) = 2^{na/m}$. Therefore, $f(2^q) = 2^{qa}$, $q \in \mathbb{Q}$. Since the numbers 2^q, $q \in \mathbb{Q}$, are dense

V. Brayman and A. Kukush, *Undergraduate Mathematics Competitions (1995–2016)*, Problem Books in Mathematics,
DOI 10.1007/978-3-319-58673-1_28

in $(0, \infty)$, due to the monotonicity of f we get $f(x) = x^a$ for all $x > 0$. Verification shows that this function satisfies the initial equation only for $a = 1$.
Answer: $f(x) = x$.

4 Let $f(x) = x^3 - 3x$. Assume that the sequence converges to ℓ. Then $f(\ell) = \ell$. Therefore, $\ell \in \{0, \pm 2\}$. In a neighborhood of the point ℓ it holds $|f'(x)| > 1$. There exists $N \in \mathbb{N}$ such that x_n lies in this neighborhood for every $n \geq N$. If $x_n \neq \ell$ then by the mean value theorem we have

$$|x_{n+1} - \ell| = |f(x_n) - f(\ell)| = |f'(\theta)| \cdot |x_n - \ell| > |x_n - \ell|.$$

Thus, the convergence holds only in the case $x_n = \ell$ for some n. Find for which values of a it is possible. If $|a| > 2$ then it is easy to verify that $|x_{n+1}| > |x_n| > 2$, $n \geq 1$. Therefore, $|a| \leq 2$, so one can set $x_1 = 2\cos\varphi$. Then $x_n = 2\cos 3^{n-1}\varphi$, $n \geq 1$. To find the answer it remains to solve the equation $2\cos 3^{n-1}\varphi = \ell$ for $\ell = 0, \pm 2$.
Answer: $a = 2\cos\dfrac{\pi k}{2 \cdot 3^{n-1}}$, $k, n \in \mathbb{N}$.

5 It is easy to check that

$$\sum_{n=1}^{\infty} \frac{d(n)}{n^2} = \left(\sum_{k=1}^{\infty} \frac{1}{k^2}\right)^2 < \left(1 + \sum_{k=2}^{\infty} \left(\frac{1}{k-1} - \frac{1}{k}\right)\right)^2 = 2^2 = 4.$$

6 Show that the wolves can catch the hare. Consider circles which are the parallels of the torus

$$K_\varphi = \left\{(x, y, z) \mid \sqrt{x^2 + y^2} = 2000 - \sqrt{2000}\cos\varphi,\ z = \sqrt{2000}\sin\varphi\right\},\quad \varphi \in (-\pi, \pi].$$

Notice that a movement along the parallel K_φ at speed 1 corresponds to the angular velocity (of movement along the parallel)

$$\omega_\varphi = \frac{1}{2000 - \sqrt{2000}\cos\varphi}.$$

At the beginning both wolves should get to the least parallel of the torus K_0. Then the wolves move along this circle in opposite directions till one of them reaches a common meridian with the hare. Then it shares a meridian with the hare for the rest of the time. The second wolf also should reach a common meridian with the hare.

Show that this is possible. Indeed, the hare cannot approach the first wolf closer than at distance 1, therefore, the hare permanently occupies the parallels with $|\varphi| \geq \varphi_0 = \frac{1}{\sqrt{2000}}$ (because a circle which is the meridian has radius $\sqrt{2000}$) and moves at the angular velocity at most ω_{φ_0}, which is less than the angular velocity of the wolves ω_0.

After both wolves get to the same meridian, they should permanently occupy a common meridian with the hare and move along the meridian in opposite directions.

At the moment when the wolves occupy a common meridian with the hare at the parallels $K_{\pm\varphi}$, $0 \le \varphi \le \pi - \varphi_0$, the hare surely occupies a parallel K_ψ, where $|\psi| \ge \varphi + \varphi_0$ (if it has not been caught yet). Hence the wolves can move along the meridian at a speed at least $v_\varphi = \sqrt{1 - \left(\frac{\omega_{\varphi+\varphi_0}}{\omega_\varphi}\right)^2}$. Since v_φ is a continuous nonnegative function on a segment, it holds $\min\limits_{0\le\varphi\le\pi-\varphi_0} v_\varphi > 0$. Therefore, the wolves will catch the hare in finite time.
Answer: yes, they can.

7 It is evident that rk $A_n \le 2$ and rk $B_n \le 1$. Thus, det $A_n = \overline{0}$ and det $B_n = \overline{0}$ for every $n > 2$.
Answer: det $A_2 = \overline{1}$, det $B_2 = \overline{1}$, det $A_n =$ det $B_n = \overline{0}$, $n > 2$.

8 Necessity:

$$\frac{1}{2} \ge |z| - \operatorname{Re} z = \sqrt{z\,\overline{z}} - \frac{z+\overline{z}}{2} = \frac{1}{2}\left(\sqrt{z} - \sqrt{\overline{z}}\right)^2.$$

One can set $u = v = \sqrt{z}$.
Sufficiency:

$$\begin{aligned}|z| - \operatorname{Re} z &= \sqrt{z\overline{z}} - \frac{z+\overline{z}}{2} = \sqrt{u\overline{u}v\overline{v}} - \frac{uv + \overline{u}\,\overline{v}}{2} \le \\ &\le \frac{u\overline{u} + v\overline{v}}{2} - \frac{uv + \overline{u}\,\overline{v}}{2} = \frac{1}{2}(u - \overline{v})(\overline{u} - v) = \frac{1}{2}|\overline{u} - v|^2 \le \frac{1}{2}.\end{aligned}$$

9 Let us select sets A_1 and A_2 element-wise. For each element k we decide whether it belongs to A_1 and/or to A_2, which gives us 4 options, so we have 4^n possible pairs of sets. Sets A_1 and A_2 will be disjoint if and only if for each element k we choose one of three options (except $k \in A_1$ and $k \in A_2$), so we have 3^n possible pairs of disjoint sets.
Answer: $(3/4)^n$.

10 For the sake of convenience add a chair on the left end of the row. Denote by 1 each person together with a free chair on the left of him/her and by 0 each other free chair. Each arrangement which satisfies the condition of the problem corresponds to a row of n ones and $N + 1 - n$ zeros, and there are $\binom{N+1-n}{n}$ such rows. Total number of arrangements in question is $\binom{N}{n}$.
Answer: $\binom{N+1-n}{n}/\binom{N}{n}$.

11 Introduce new variables $t = xy$, $y = y$ in the second integral, and get

$$\int_0^1\int_0^1 (xy)^{xy}\,dx\,dy = \int_0^1\left(\int_t^1 t^t\frac{1}{y}dy\right)dt = \int_0^1 t^t(-\ln t)\,dt = \int_0^1 t^t\,dt.$$

The latter equality holds because $\int_0^1 t^t(1+\ln t)\,dt = t^t\big|_0^1 = 0$.
Answer: the integrals are equal.

12 A solution is similar to the solution of Problem 4.
Answer: $a = 2\sin\frac{\pi k}{3^{n-1}}$, $a = 2\sin\frac{\pi(1+2k)}{4\cdot 3^{n-1}}$, $k \in \mathbb{Z}$, $n \in \mathbb{N}$.

13 Unit element is self-double because for each $u \neq e$ it holds $uu^{-1} = u^{-1}u = e$. If x has order $n \geq 3$ then $x^{n-1}x^2 = x^2x^{n-1} = x$, so x is self-double.

Let x be not self-double, hence $x^2 = e$, and $x^{-1} = x$. Consider orbits of the mapping $u \mapsto xu^{-1}$, $G \to G$. We have

$$u \mapsto xu^{-1} \mapsto xux \mapsto u^{-1}x \mapsto u.$$

If $u \neq e$ and $u \neq x$ then $u \neq xux$, otherwise for $v = xu^{-1} \neq e$ we get $vu = x$ and $uv = xuxxu^{-1} = x$, so x is self-double. Therefore, all the orbits except $e \mapsto x \mapsto e$ have length 4. Hence the number of elements in the group equals $4k - 2$, where k is the number of orbits.

14 Let φ be a required homomorphism, $A = \left(\begin{smallmatrix}0&1\\0&0\end{smallmatrix}\right)$, $B = \left(\begin{smallmatrix}0&0\\1&0\end{smallmatrix}\right)$, O_2, I_2, O_3, and I_3 be zero and identity matrices of size 2×2 and 3×3, respectively. Since $A^2 = B^2 = O_2$, $(A+B)^2 = I_2$, we have $(\varphi(A))^2 = (\varphi(B))^2 = O_3$, $(\varphi(A)+\varphi(B))^2 = I_3$. Then it is easy to verify that $\operatorname{rk}\varphi(A) \leq 1$ and $\operatorname{rk}\varphi(B) \leq 1$ (to do this, it suffices to consider the Jordan normal forms of the matrices). Hence $\operatorname{rk}(\varphi(A)+\varphi(B)) \leq 2$, and the equality $(\varphi(A)+\varphi(B))^2 = I_3$ is impossible, a contradiction.
Answer: there is no such homomorphism.

15 Since $d\left(\frac{x^4}{4}+\frac{y^2}{2}\right) = x^3dx + ydy = 0$, for every solution it holds $\frac{x^4}{4}+\frac{y^2}{2} =$ const. Stationary solutions have a form $x = c$, $y = c$, where $c \in \mathbb{R}$ is fixed. Each other solution starts from the point $(-c, -c)$ and comes to the point (c, c), where c is such that $c \geq 0$ and $\frac{x^4}{4}+\frac{y^2}{2} = \frac{c^4}{4}+\frac{c^2}{2}$. Therefore, this solution is not periodic. Hence there is no periodic solution, which differs from constant function.

16
(a) To be specific let us assume that the null solution $x(t, 0, \vec{0})$, $t \geq 0$, is stable in the sense of Lyapunov. Consider an arbitrary $\varepsilon > 0$ and $\delta = \delta(\varepsilon, 0)$ from the definition of stability. Since the system is autonomous, for every t_0 it holds

$$x(t, t_0, \vec{0}) = x(t - t_0, 0, \vec{0}),\ t \geq t_0,$$

therefore $\|x_0\| < \delta$ implies that

$$\|x(t, t_0, x_0)\| = \|x(t - t_0, 0, x_0)\| < \varepsilon,\ t \geq t_0.$$

(b) Choose $R > 0$ such that f is a Lipschitz function on the set

$$B_R = \{x \in \mathbb{R}^n \mid \|x\| \leq R\}$$

and $r > 0$ such that $x_0 \in B_r$ implies $x(t, t_0, x_0) \in B_R$, $t \geq t_0$. For each $x_0 \in B_r$ it holds $\|x(t, t_0, x_0)\| \to 0$, as $t \to \infty$.

Show that $\lim_{t\to\infty} \|x(t, t_0, x_0)\| = 0$ uniformly in $x_0 \in B_r$. Fix an arbitrary $\varepsilon > 0$. As a result of (a), there exists $\delta > 0$ such that for every $t_0 \in \mathbb{R}$ the inequality $\|x_0\| < \delta$ implies that $\|x(t, t_0, x_0)\| < \varepsilon, t \geq t_0$. For each $y \in B_R$, there exists $T_y > t_0$ such that $\|x(T_y, t_0, y)\| < \delta/2$. Since $\|x(T_y, t_0, y)\|$ is continuous in y, there exists a neighborhood $B(y)$ of the point y such that for $z \in B(y) \bigcap B_r$ we have $\|x(T_y, t_0, z)\| < \delta$. Hence for $z \in B(y) \bigcap B_r$ and $t \geq T_y$ we have

$$\|x(t, t_0, z)\| = \|x(t - T_y, T_y, x(T_y, t_0, z))\| < \varepsilon.$$

The sets $B(y)$ form an open cover of the compact set B_r. Select a finite subcover $B(y_1), \ldots, B(y_k)$. Let $T(\varepsilon) = \max(T_{y_1}, \ldots, T_{y_k})$. Then for $t \geq T(\varepsilon)$ and for all $x_0 \in B_r$ it holds $\|x(t, t_0, x_0)\| < \varepsilon$, which proves the uniform convergence.

17
(a) Let μ be a measure on $\mathbb{N}$ such that each positive integer has measure 1. Take arbitrary $b > a \geq 1$. By change of variables $nx \to x$ and Fubini's theorem we have

$$\int_{[a,b]} \left(\sum_{n=1}^{\infty} f(nx) \right) d\lambda(x) = \int_{[a,b]} \int_{\mathbb{N}} f(nx)\, d\mu(n)\, d\lambda(x) =$$
$$= \int_{\mathbb{N}} \int_{[a,b]} f(nx)\, d\lambda(x)\, d\mu(n) = \sum_{n=1}^{\infty} \tfrac{1}{n} \int_{na}^{nb} f(x)\, d\lambda(x) =$$
$$= \int_{a}^{\infty} \sum_{\{n:\, na \leq x \leq nb\}} \tfrac{1}{n} f(x)\, d\lambda(x) \leq \tfrac{b}{a} \int_{a}^{\infty} f(x)\, d\lambda(x) < +\infty,$$

because the sum $\sum_{\{n:\, na \leq x \leq nb\}} \frac{1}{n}$ contains at most $\frac{x}{a}$ terms, each of which does not exceed $\frac{b}{x}$. Fubini's theorem implies the required statement.
(b) For every $\varepsilon > 0$, choose $b > a > 0$ large enough such that $\int_a^\infty f(x)\, d\lambda(x) < \varepsilon$ and $\frac{a}{b} \int_1^\infty f(x)\, d\lambda(x) < \varepsilon$. Then for each $T > b$ it holds

$$\frac{1}{T} \int_{1}^{T} x f(x)\, d\lambda(x) = \frac{1}{T} \int_{1}^{a} x f(x)\, d\lambda(x) + \frac{1}{T} \int_{a}^{b} x f(x)\, d\lambda(x) \leq$$
$$\leq \frac{a}{T} \int_{1}^{a} f(x)\, d\lambda(x) + \frac{T}{T} \int_{a}^{b} f(x)\, d\lambda(x) \leq \varepsilon + \varepsilon = 2\varepsilon.$$

18 Notice that $x + (\xi - x)_+ = \xi + (x - \xi)_+ = \max(x, \xi)$, $x \geq 0$. Hence

$$x + f(x) = \mathsf{E}(\xi + (x - \xi)_+) = \mathsf{E}\xi + \int_0^x (x - y)\, d\mathsf{P}(\xi < y) =$$
$$= \mathsf{E}\xi + \int_0^x \mathsf{P}(\xi < y)\, dy.$$

Therefore, there exists the derivative $(f(x)+x)' = \mathsf{P}(\xi < x)$ almost everywhere, and

$$\mathsf{E}e^{\xi} = \int_0^{+\infty} e^x d\mathsf{P}(\xi < x) = \int_0^{+\infty} e^x d\left((f(x)+x)'\right).$$

19 Denote by $\nu(x)$ the number of passengers at the bus stop at a moment $x \geq 0$. If the passengers come to the stop at moments

$$t_1 \leq t_2 \leq \ldots \leq t_{\nu(t)},$$

then the waiting time of all the passengers equals

$$\sum_{k=1}^{\nu(t)-1} k(t_{k+1} - t_k) + \nu(t)(t - t_{\nu(t)}) = \int_0^t \nu(x)\,dx.$$

By Fubini's theorem we find

$$\mathsf{E}\int_0^t \nu(x)\,dx = \int_0^t \mathsf{E}\nu(x)\,dx = \int_0^t \lambda x\,dx = \frac{1}{2}\lambda t^2.$$

Answer: $\frac{1}{2}\lambda t^2$.

2001

1 On a circle of the unit length mark the points $k\pi$, $1 \le k \le n_0$. For some $1 \le k_1 < k_2 \le n_0$ the distance along the circle between the points $k_1\pi$, $k_2\pi$ does not exceed $\frac{1}{n_0}$, hence for $k_0 = k_2 - k_1$, $k_0 \le n_0$ and for some $l_0 \in \mathbb{Z}$ it holds $|k_0\pi - l_0| \le \frac{1}{n_0}$. Thus,

$$|\sin l_0| = |\sin k_0\pi - \sin l_0| \le |k_0\pi - l_0| \le \tfrac{1}{n_0}.$$

Since $|l_0| < |k_0\pi| + 1 < 4n_0 + 1$, we have $|l_0 \sin l_0| < 5$. It is evident that $l_0 \to \infty$, as $n_0 \to \infty$. Therefore, $|n \sin n| \not\to +\infty$.
Answer: no, it is not.

2
(a) Suppose that for all x it holds $f(x)f''(x) + 2(f'(x))^2 < 0$. Then the function f is not constant at any interval. If $f(x_0) = 0$ then one can set $\theta = x_0$. Suppose that $f(x) \ne 0$, $x \in \mathbb{R}$. Then f does not change the sign, e.g., $f(x) > 0$, $x \in \mathbb{R}$. But in that case $(f^3)'' = 3f\left(ff'' + 2(f')^2\right) < 0$, so f^3 is concave, which is impossible for a positive nonconstant function.
(b) Show that, e.g., the function $G(t) = |t|^3$ has required property. It holds

$$(G(f))'' = 3|f|\left(ff'' + 2(f')^2\right).$$

The necessity is evident.

Prove the sufficiency. Suppose that $G(f(x))$ is convex and show that

$$|f(x)|\left(f(x)f''(x) + 2(f'(x))^2\right) \ge 0, \ x \in \mathbb{R}.$$

If $f(x) \ne 0$ for all $x \in \mathbb{R}$, then the required inequality holds. Otherwise the convexity of $|f(x)|^3$ implies that the set of zeros of $f(x)$ is convex, so this set is either a segment or a single point. If this set is a point then the inequality holds at this point due to

V. Brayman and A. Kukush, *Undergraduate Mathematics Competitions (1995–2016)*, Problem Books in Mathematics,
DOI 10.1007/978-3-319-58673-1_29

continuity of f' and f''. If this set is a segment then $f' = f'' = 0$ on the segment, and the inequality is evident.

3 Introduce $b_n = \ln \frac{2a_n}{3}$. We have to show that $\lim_{n\to\infty} b_n \in \left(\frac{1}{4}, \frac{1}{2}\right)$. It is easy to check that $\frac{x}{2} < \ln(1+x) < x$, $x \in (0, 1)$. Thus,

$$\frac{1}{4} = \sum_{n\geq 2} 2^{-n-1} < \lim_{n\to\infty} b_n = \sum_{n\geq 2} \ln(1+2^{-n}) < \sum_{n\geq 2} 2^{-n} = \frac{1}{2},$$

which finishes the proof.

4 Let $(a_1, a_2, \ldots, a_n)$ be a solution. Suppose that there are exactly r distinct numbers among $a_1, \ldots, a_n$. We may assume the first r numbers are distinct. Let a_j repeats m_j times, $1 \leq j \leq r$. After reduction of similar terms, we get

$$m_1 a_1^k + m_2 a_2^k + \ldots + m_r a_r^k = 0, \quad k = 1, 2, \ldots, n.$$

Denote $y_1 = m_1 a_1, \ \ldots, \ y_r = m_r a_r$. We have

$$a_1^{k-1} y_1 + a_2^{k-1} y_2 + \ldots + a_r^{k-1} y_r = 0, \quad k = 1, 2, \ldots, n.$$

The first r equations form a system of linear equations with variables $y_1, y_2, \ldots, y_r$. Its determinant is the Vandermonde one for the numbers $a_1, a_2, \ldots, a_r$. These numbers are distinct, and hence the determinant is nonzero, so the system has a unique solution $y_1 = y_2 = \ldots = y_r = 0$. Thus, $a_1 = a_2 = \ldots = a_r = 0$, therefore, $r = 1$ and the system has only zero solution.
Answer: (0,…,0).

5 The columns of the matrix $\binom{B}{D}$ are linear combinations of the columns of the matrix $\binom{A}{C}$. Hence $B = AX$, and $D = CX$ for some matrix X. Thus, $X = A^{-1}B$ and $D = CX = CA^{-1}B$.

6 Write out all the couples consisting of a permutation and an element which is a fixed point of this permutation. The number of such couples equals $\sum_{k=1}^{n} k \cdot b(n, k)$. On the other hand, for each of n elements, there exist $(n-1)!$ permutations for which this element is a fixed point. Thus, the total number of the couples equals $n \cdot (n-1)! = n!$
Answer: $n!$

8 A linear combination of bounded solutions is a bounded solution as well, and hence B is a subspace. Let $x(t)$ be a bounded solution to $(*)$. Decompose $x(0) = a_1 + a_2$, where $a_1 \in B$ and $a_2 \in B^{\perp}$. Denote by $x_1(t)$ a solution to the homogeneous equation $\frac{dx}{dt} = A(t)x$ with initial condition $x_1(0) = a_1$, which is bounded by definition of the subspace B. Then $x_2(t) = x(t) - x_1(t)$ is a bounded solution to the equation $(*)$, and $x_2(0) = x(0) - a_1 = a_2 \in B^{\perp}$. To prove the uniqueness consider the difference of two bounded solutions to (*) with initial conditions in $B^{\perp}$. It is a bounded solution to the homogeneous equation which starts in $B^{\perp}$, i.e., zero solution.

9 A solution is similar to the solution of Problem 6.
Answer: 1.

10 Denote by $g(r)$ the radius of a circle which is the image of a circle with center at the origin and radius r. By the maximum principle the function $g(r)$ is monotone. Therefore, either f is an analytic function on $\mathbb{C}$ or it has a pole at zero.

Let f be analytic on $\mathbb{C}$. If $f(0) \neq 0$ then the function $\dfrac{1}{f(z)}$ is bounded and analytic on $\mathbb{C}$, whence $f = \text{const}$. If $f(0) = 0$, denote by r an order of zero at point 0. The function $\dfrac{f(z)}{z^r}$ differs from 0 at zero, so by previous reasonings it is constant, i.e., $f(z) = Cz^r$, $r \in \mathbb{N}$. In a similar way we get that if the function f has a pole at zero then $f(z) = Cz^r$, where $-r \in \mathbb{N}$.
Answer: $f(z) = Cz^r, r \in \mathbb{Z}$.

11 Show that a convex set in $\mathbb{R}^n$ which contains a cone coincides with $\mathbb{R}^n$. It suffices to consider the cone $\{(x_1, \dots, x_n) : x_1^2 + \dots + x_{n-1}^2 \leq rx_n^2\}$. For each point $(y_1, \dots, y_n) \in \mathbb{R}^n$, there exists N large enough such that the points $(y_1, \dots, y_{n-1}, -N)$ and $(y_1, \dots, y_{n-1}, N)$ belong to the cone and therefore, to the convex set. Since the set is convex, it contains the point $(y_1, \dots, y_{n-1}, y_n)$ as well. Thus, the set coincides with $\mathbb{R}^n$.

A projection of a convex set is a convex set, and hence the statement of the problem is equivalent to the following: if A is an unbounded convex set in $\mathbb{R}^n$ and $A \neq \mathbb{R}^n$, then there exists a two-dimensional subspace $B \subset \mathbb{R}^n$ such that the projection of A onto B does not coincide with B. By the Separating Axis Theorem, there exists a line ℓ in $\mathbb{R}^n$ such that the projection of A onto ℓ does not coincide with ℓ. Then an arbitrary two-dimensional subspace $B \subset \mathbb{R}^n$ for which $\ell \subset B$ has the required property.

Remark 1 The statement of the problem is still correct for *one-sided* cones, i.e., sets which are obtained by shift and rotation of the set

$$\{(x_1, \dots, x_n) : x_1^2 + \dots + x_{n-1}^2 \leq rx_n^2, \ x_n \geq 0\}$$

for some $r > 0$. Since A is an unbounded convex set, it contains some ray ℓ. Let O be the beginning of the ray ℓ and γ be a hyperplane orthogonal to ℓ, which intersects ℓ at a point different from O. Introduce a set G of all intersection points of γ and the rays which start at O and are contained in A. This set is convex and does not contain any $(n-1)$-dimensional ball in γ because A does not contain a cone. Hence G is contained in some $(n-2)$-dimensional subspace. The two-dimensional subspace B, which is orthogonal to this subspace, has the required property. Indeed, the projection of A onto B contains ℓ. Show that it does not contain any other ray. Assume that it contains some ray m. Consider the subset of A which is projected onto m. This set is unbounded and convex; therefore, it contains some ray. But the ray cannot intersect γ at a point from G, a contradiction.

12 By the strong law of large numbers

$$\frac{1}{n}\sum_{k=1}^{n}\gamma_n^2 \xrightarrow{\mathsf{P}1} \mathsf{E}\gamma_1^2 = 1, \text{ as } n \to \infty.$$

Show that $\frac{1}{\ln n}\max\limits_{1\le k\le n}\gamma_k^2 \xrightarrow{\mathsf{P}} 2$, as $n \to \infty$. Consider

$$F_n(x) = \mathsf{P}\left(\frac{1}{\ln n}\max_{1\le k\le n}\gamma_k^2 \le x\right),\ x \ge 0.$$

It suffices to prove that if $x < 2$ then $F_n(x) \to 0$, as $n \to \infty$, and if $x \ge 2$ then $F_n(x) \to 1$, as $n \to \infty$.

Random variables $\gamma_1, \dots, \gamma_n$ are independent, hence

$$F_n(x) = \mathsf{P}\left(\gamma_k^2 \le x\ln n,\ 1 \le k \le n\right) = \mathsf{P}^n(\gamma_1^2 \le x\ln n) = \left(2\int_0^{\sqrt{x\ln n}} \frac{e^{-t^2/2}}{\sqrt{2\pi}}dt\right)^n.$$

Show that

$$\frac{1}{2t}e^{-t^2/2} < \int_t^{+\infty} e^{-s^2/2}ds < \frac{1}{t}e^{-t^2/2},\quad t \ge 1. \tag{$*$}$$

We have

$$\int_t^{+\infty} e^{-s^2/2}ds < \int_t^{+\infty} \frac{s}{t}e^{-s^2/2}ds = \frac{1}{t}e^{-t^2/2}.$$

To prove the left-hand side inequality notice that the function $y = e^{-x^2/2}$ is convex for $x \ge 1$ ($y''(x) = (x^2-1)e^{-x^2/2} \ge 0$), hence the area under the graph $y = e^{-x^2/2}$ exceeds the area of the triangle, which is bounded by the straight lines $x = t$ and $y = 0$ and the tangent to the graph at point t, that is $y = -te^{-t^2/2}(x-t) + e^{-t^2/2}$.

For $0 < x < 2-\varepsilon$, $0 < \varepsilon < 1$, and for n large enough ($*$) implies that

$$F_n(x) \le \left(1 - 2\cdot\frac{e^{-x\ln n/2}}{2\sqrt{2\pi x\ln n}}\right)^n <$$
$$< \left(1 - \frac{1}{\sqrt{4\pi\ln n}\,n^{1-\varepsilon/2}}\right)^n < \left(1 - \frac{1}{n^{1-\varepsilon/3}}\right)^n \to 0,\text{ as } n \to \infty.$$

For $x \geq 2$ due to $(*)$ we have

$$1 \geq F_n(x) \geq \left(1 - 2 \cdot \frac{e^{-x \ln n/2}}{\sqrt{2\pi x \ln n}}\right)^n \geq$$
$$\geq \left(1 - \sqrt{\frac{1}{\pi \ln n}} \cdot \frac{1}{n}\right)^n \geq 1 - \sqrt{\frac{1}{\pi \ln n}} \to 1, \text{ as } n \to \infty.$$

Hence $\frac{1}{\ln n} \max\limits_{1 \leq k \leq n} \gamma_k^2 \xrightarrow{\mathsf{P}} 2$, as $n \to \infty$, therefore,

$$\frac{\max\limits_{1 \leq k \leq n} \gamma_k^2}{\sum\limits_{k=1}^{n} \gamma_k^2} : \frac{\ln n}{n} \xrightarrow{\mathsf{P}} 2, \text{ as } n \to \infty.$$

2002

1 Enumerate all the rational numbers $\{r_i, i \geq 1\}$. Introduce sets

$$A_i^+ = \{(u, v) : u > r_i > v\} \quad \text{and} \quad A_i^- = \{(u, v) : u < r_i < v\}.$$

There are no x, y, z such that both points (x, y) and (y, z) belong to one of the sets A_i^+ or A_i^-. It is evident that

$$\bigcup_{i\geq 1} \left(A_i^+ \cup A_i^-\right) \cup A_0 = \mathbb{R}^2,$$

where $A_0 = \{(x, x) : x \in \mathbb{R}\}$. Renumber the sets A_0 and A_i^+, A_i^-, $i \geq 1$, as B_1, $B_2, \ldots$, and put $F(x, y) = \min\{k \in \mathbb{N} : (x, y) \in B_k\}$. For distinct points (x, y) and (y, z) it holds $F(x, y) \neq F(y, z)$, so the function F has required property.
Answer: a function does exist.

2 For $x = \pi$ it holds $y = 1 + \frac{1}{a}$. If $1 \leq a \leq 2.5$ then $1.4 \leq y \leq 2$. Hence the distance from the point $M(\pi, 1.7)$ to each of the graphs in question does not exceed $0.3 < 0.4$.

3 Introduce a function $g(x) = f(x) \cdot e^{-\arctan x}$, $x \in [-1, 1]$. Since $g(-1) = g(1) = 0$, by Rolle's theorem there exists a point $x \in (-1, 1)$ such that $g'(x) = 0$, i.e.,

$$f'(x)e^{-\arctan x} - f(x)e^{-\arctan x} \cdot \frac{1}{1 + x^2} = 0, \quad f(x) = (1 + x^2)f'(x).$$

4 Let J be the matrix with all entries equal to 1 and I be the identity matrix. We have $I = J - A - A^{\mathrm{T}}$. It holds $\operatorname{rk} J = 1$. Assume that $\operatorname{rk} A < n - 1$. Then $\operatorname{rk} A + \operatorname{rk} J < n$, so there exists a nonzero vector v, for which $Av = 0$ and $Jv = 0$. It follows that

V. Brayman and A. Kukush, *Undergraduate Mathematics Competitions (1995–2016)*, Problem Books in Mathematics,
DOI 10.1007/978-3-319-58673-1_30

$$(v, v) = v^{\mathrm{T}}(J - A - A^{\mathrm{T}})v = v^{\mathrm{T}} J v - v^{\mathrm{T}} A v - (Av)^{\mathrm{T}} v = 0,$$

a contradiction.

5 Use the change of variables $y = \frac{\pi}{2} - x$. We get

$$\int_0^{\frac{\pi}{2}} \frac{(\cos x)^{\sin x}}{(\cos x)^{\sin x} + (\sin x)^{\cos x}} dx = \int_0^{\frac{\pi}{2}} \frac{(\sin y)^{\cos y}}{(\cos y)^{\sin y} + (\sin y)^{\cos y}} dy,$$

whence $2I = \int_0^{\frac{\pi}{2}} 1 dx = \frac{\pi}{2}$. Therefore, $I = \frac{\pi}{4} < 1$.

6 Each linear operator φ on $M_n(\mathbb{R})$ is determined by a collection of numbers φ_{ijkl}, $1 \le i, j, k, l \le n$ as follows: if $A = (a_{ij})$ then $\varphi(A) = (a_{ij}^{\varphi})$, where

$$a_{ij}^{\varphi} = \sum_{k,l=1}^{n} \varphi_{ijkl} a_{kl}.$$

The condition $\varphi(A^{\mathrm{T}}) = (\varphi(A))^{\mathrm{T}}$ means that $\varphi_{ijkl} = \varphi_{jilk}$ for all i, j, k, l. Hence in order to determine φ one has choose the coefficients φ_{ijkl} for $i < j$ and arbitrary k, l, and also for $i = j$ and $k \le l$. Thus, one has to cast

$$\frac{n(n-1)}{2} \cdot n^2 + n \cdot \frac{n(n+1)}{2} = \frac{n^2(n^2+1)}{2}$$

numbers. This is the dimension of the subspace in question.
Answer: $\frac{n^2(n^2+1)}{2}$.

7 By induction on k prove that $\dfrac{a_k}{e}$ is an integer for each $k \ge 1$. It holds $a_1 = e$. Assume that $a_k = e \cdot b_k$, $k \le n$, where b_k are some integers. Then

$$a_{n+1} = \sum_{j=1}^{\infty} \frac{j^{n+1}}{j!} = \sum_{j=1}^{\infty} \frac{j^n}{(j-1)!} = \sum_{j=0}^{\infty} \frac{(j+1)^n}{j!} =$$

$$= \sum_{j=1}^{\infty} \sum_{m=0}^{n} \binom{n}{m} \frac{j^m}{j!} = \sum_{m=0}^{n} \binom{n}{m} \cdot \sum_{j=0}^{\infty} \frac{j^m}{j!} = e + \sum_{m=1}^{n} \binom{n}{m} \cdot \sum_{j=1}^{\infty} \frac{j^m}{j!} =$$

$$= e + \sum_{m=1}^{n} \binom{n}{m} a_m = e\left(1 + \sum_{m=1}^{n} \binom{n}{m} b_m\right),$$

so $\dfrac{a_{n+1}}{e}$ is an integer, and the statement is proved. Hence $a_k \notin \mathbb{Q}$, $k \ge 1$.

8 First show that $f \in C^{(\infty)}(\mathbb{R})$. To do this integrate the left-hand and right-hand sides of the equation over $y \in [0,1]$ set $z = 0$, and change variables in the integrals in the right-hand side:

$$\int_0^1 (f(x) + f(y) + f(0))\,dy =$$
$$= \int_0^1 \left(f\left(\tfrac{3}{7}x + \tfrac{6}{7}y\right) + f\left(\tfrac{6}{7}x - \tfrac{2}{7}y\right) + f\left(-\tfrac{2}{7}x + \tfrac{3}{7}y\right)\right) dy,$$

$$f(x) + \int_0^1 f(y)dy + f(0) =$$
$$= \int_{\frac{3}{7}x}^{\frac{3}{7}x+1} f\left(\tfrac{6}{7}y\right) dy + \int_{\frac{6}{7}x}^{\frac{6}{7}x+1} f\left(-\tfrac{2}{7}y\right) dy + \int_{-\frac{2}{7}x}^{-\frac{2}{7}x+1} f\left(\tfrac{3}{7}y\right) dy.$$

Since $f \in C(\mathbb{R})$, by the theorem on differentiating with respect to the limit of integration the right-hand side of the latter equality belongs to $C^{(1)}(\mathbb{R})$, hence $f \in C^{(1)}(\mathbb{R})$. Then the same equality implies that $f \in C^{(2)}(\mathbb{R})$, etc. Differentiate the initial equation three times with respect to x, and set $x = y = z$. We get

$$f'''(x) = \left(\left(\tfrac{3}{7}\right)^3 + \left(\tfrac{6}{7}\right)^3 + \left(-\tfrac{2}{7}\right)^3\right) f'''(x),$$

whence $f'''(x) \equiv 0$, $x \in \mathbb{R}$. Thus, f is a polynomial of at most second degree. It is easy to verify that the functions 1, x, and x^2 satisfy the condition of the problem. Therefore, every polynomial of at most second degree satisfies the condition.
Answer: $f(x) = ax^2 + bx + c$, $a, b, c \in \mathbb{R}$.

9 Let $A \subset [0,1]$ be the Cantor set, i.e., the set of the numbers $\sum\limits_{k=1}^{\infty} \frac{\alpha_k}{3^k}$, where $\alpha_k \in \{0, 2\}$ for all $k \geq 1$. Put

$$f\left(\sum_{k=1}^{\infty} \frac{\alpha_k}{3^k}\right) = \sum_{k=1}^{\infty} \frac{\alpha_k}{3^{3k+1}}.$$

Let $a_1 = \sum\limits_{k=1}^{\infty} \frac{\alpha_k^{(1)}}{3^k}$ and $a_2 = \sum\limits_{k=1}^{\infty} \frac{\alpha_k^{(2)}}{3^k}$. Assume that $a_1 < a_2$ and the first difference between the ternary notations of the numbers a_1 and a_2 occurs at nth digit after point, i.e., $\alpha_1^{(1)} = \alpha_1^{(2)}, \ldots, \alpha_{n-1}^{(1)} = \alpha_{n-1}^{(2)}, \alpha_n^{(1)} = 0$, and $\alpha_n^{(2)} = 2$. Then

$$a_2 - a_1 \geq \frac{2}{3^n} - \sum_{k=n+1}^{\infty} \frac{2}{3^k} = \frac{1}{3^n}$$

and

$$0 < f(a_2) - f(a_1) \le \sum_{k=n}^{\infty} \frac{2}{3^{3k+1}} = \frac{9}{13} \cdot \frac{1}{3^{3n}} < |a_2 - a_1|^3.$$

Hence the function f satisfies the condition $|f(a_2) - f(a_1)| < |a_2 - a_1|^3$. Since f is injective and the set A is uncountable, the range of f is uncountable as well.

10 Take any vertex A of the prismatoid and divide each face, which does not contain the vertex A, into triangles by its diagonals. The prismatoid can be dissected into triangular pyramids with vertex A, the bases of which are these triangles. The constructed pyramids are prismatoids with the same bases as initial prismatoid, because the dissection does not create new vertices. It remains to check the the statement of the problem holds true for a prismatoid, which is a triangular pyramid, in two cases: (a) three vertices belong to one of the two parallel planes and another plane contains the fourth vertex, and (b) each of the two parallel planes contains two vertices.

11 Consider sets $A_{n,k} = \left\{x \in \Omega \mid \frac{k}{n} \le \xi(x) < \frac{k+1}{n}\right\} \in \mathscr{F}$, where $n \ge 1$, $k \in \mathbb{Z}$. Then for every set $A \in \mathscr{F}$ it holds

$$\frac{k}{n}\mathsf{P}(A \cap A_{n,k}) \le \omega(A \cap A_{n,k}) \le \frac{k+1}{n}\mathsf{P}(A \cap A_{n,k})$$

and

$$\frac{k}{n}\mathsf{P}(A \cap A_{n,k}) \le \int_{A \cap A_{n,k}} \xi(x) d\mathsf{P}(x) \le \frac{k+1}{n}\mathsf{P}(A \cap A_{n,k}).$$

Thus,

$$\left|\omega(A) - \int_A \xi(x) d\mathsf{P}(x)\right| \le \sum_{k \in \mathbb{Z}} \left|\omega(A \cap A_{n,k}) - \int_{A \cap A_{n,k}} \xi(x) d\mathsf{P}(x)\right| \le$$

$$\le \frac{1}{n} \cdot \sum_{k \in \mathbb{Z}} \mathsf{P}(A \cap A_{n,k}) \le \frac{1}{n} \to 0, \text{ as } n \to \infty,$$

whence $\omega(A) = \int_A \xi(x) d\mathsf{P}(x)$, $A \in \mathscr{F}$.

12 Since the functions $f_n(x)$ are 2π-periodic, it suffices to deal with the interval $[0, 2\pi]$. Moreover, for each x there exists $x_0 \in \left[0, \frac{\pi}{2}\right]$ such that $\sin x_0 = |\sin x|$ and $\cos x_0 = |\cos x|$, hence $f_n(x_0) \ge f_n(x)$. Therefore, it suffices to consider $x \in \left[0, \frac{\pi}{2}\right]$. Finally, the graph $y = f_n(x)$ has an axis of symmetry $x = \frac{\pi}{4}$. Therefore, we can consider $x \in \left[0, \frac{\pi}{4}\right]$. For each n, the global maximum of $f_n(x)$ is attained at some point $x_n \in \left[0, \frac{\pi}{4}\right]$. A point of global maximum satisfies the conditions

$$f'(x) = \ln n \left(n^{\sin x} \cdot \cos x - n^{\cos x} \cdot \sin x\right) = 0$$

and

$$f''(x) = \ln^2 n\left(n^{\sin x}\cdot\cos^2 x + n^{\cos x}\cdot\sin^2 x\right) - \ln n\left(n^{\sin x}\cdot\sin x + n^{\cos x}\cdot\cos x\right) \le 0.$$

Since $n^{\sin x}\cdot\cos x = n^{\cos x}\cdot\sin x$, in case $\cos x \ge \sin x > \frac{1}{\ln n}$ it holds

$$f''(x) = \ln n\left(n^{\cos x}\cdot\cos x\,(\ln n\cdot\sin x - 1) + n^{\sin x}\cdot\sin x\,(\ln n\cdot\cos x - 1)\right) > 0,$$

hence the point of global maximum $x_n \in \left[0, \frac{\pi}{4}\right]$ satisfies the condition $\sin x_n \le \frac{1}{\ln n}$, so $x_n < \arcsin\frac{1}{\ln n} \to 0$, as $n \to \infty$.

13 Verify that

$$P_{U^{\mathrm{T}}L}(U^{-1}a) = P_{U^{\mathrm{T}}L}(U^{-1}P_La),\ \text{ that is }\ (U^{-1}a - U^{-1}P_La)\perp U^{\mathrm{T}}L.$$

Indeed, for each $\ell\in L$ it holds

$$\left(U^{-1}a - U^{-1}P_La, U^{\mathrm{T}}\ell\right) = \left(UU^{-1}(a-P_La),\ell\right) = (a-P_La,\ell) = 0$$

by the definition of P_L. Hence

$$\|P_{U^{\mathrm{T}}L}(U^{-1}a)\| = \|P_{U^{\mathrm{T}}L}(U^{-1}P_La)\| \le \|U^{-1}P_La\| \le \|U^{-1}\|\cdot\|P_La\|.$$

14 Let $z(t) = x(t) + iy(t)$, $x(t), y(t)\in\mathbb{R}$. Then $x(t)$, $y(t)$ satisfy the system of equations

$$\begin{cases}\frac{dx}{dt} = -yf(x+iy),\\ \frac{dy}{dt} = xf(x+iy),\end{cases}$$

whence $\frac{d}{dt}(x^2+y^2) = x\frac{dx}{dt} + y\frac{dy}{dt} = 0$, and $x^2(t)+y^2(t) = |z(t)|^2 = \text{const}$. So it is convenient to use the polar coordinates. Let $z(t) = |z(0)|\cdot e^{i\varphi(t)} = re^{i\varphi(t)}$. We have $\frac{dz}{dt} = ire^{i\varphi}\frac{d\varphi}{dt} = izf(z) = ire^{i\varphi}f(re^{i\varphi})$, so $\frac{d\varphi}{dt} = f(re^{i\varphi})$. Solve the differential equation and get

$$\int\frac{d\varphi}{f\left(re^{i\varphi}\right)} = \int dt,$$

whence $F_r(2\pi) - F_r(0) = t_r$, where $F_r(\varphi)$ is the primitive of $\frac{1}{f\left(re^{i\varphi}\right)}$, and t_r is the when the solution which comes from the point $r = r\cdot e^{i\cdot 0}$ returns to the point $r = r\cdot e^{i\cdot 2\pi}$. Notice that the function $\frac{1}{f\left(re^{i\varphi}\right)}$ is integrable, because for each r_0 the continuous function f is bounded from below on the compact set $\{r = r_0,\ 0\le\varphi\le 2\pi\}$ by some constant $M_{r_0} > 0$. By the theorem on continuity of an integral with respect to a parameter, the function F_r is continuous in r, hence t_r is a continuous function of r as well. Since the solution $z(t) = re^{i\varphi(t)}$, $z(0) = r$ has a period t_r, it remains to notice that

$$t_r \geq 2\pi \cdot \left(\max_{\varphi} f\left(re^{i\varphi}\right) \right)^{-1} \to \infty, \text{ as } r \to 0+,$$

$$t_r \leq 2\pi \left(\min_{\varphi} f\left(re^{i\varphi}\right) \right)^{-1} \to 0, \text{ as } r \to \infty,$$

hence the continuity of t_r implies that for each $T > 0$ there exists R, for which $t_R = T$. The solution to the equation $\frac{dz}{dt} = izf(z)$ with initial condition $z(0) = R$ is periodic with a period of T.

2003

1 Decompose the fraction $\dfrac{9n+4}{n(3n+1)(3n+2)}$ into a sum of partial fractions:

$$\sum_{n=1}^{N}\frac{9n+4}{n(3n+1)(3n+2)}=\sum_{n=1}^{N}\left(\frac{2}{n}-\frac{3}{3n+1}-\frac{3}{3n+2}\right)=$$

$$=\sum_{n=1}^{N}\left(\frac{3}{n}-\frac{3}{3n}-\frac{3}{3n+1}-\frac{3}{3n+2}\right)=3\left(\sum_{n=1}^{N}\frac{1}{n}-\sum_{n=3}^{3N+2}\frac{1}{n}\right)=$$

$$=3\left(1+\frac{1}{2}-\sum_{n=N+1}^{3N+2}\frac{1}{n}\right)=\frac{9}{2}-3\sum_{k=1}^{2N+2}\frac{1}{N+k}=$$

$$=\frac{9}{2}-3\cdot\frac{1}{N}\sum_{k=1}^{2N+2}\frac{1}{1+\frac{k}{N}}\to\frac{9}{2}-3\int_0^2\frac{dx}{1+x}=\frac{9}{2}-3\ln 3,\ \text{as } N\to\infty.$$

Answer: $\frac{9}{2}-3\ln 3$.

2 Estimate $\max\limits_{1\le n\le N}\left\{\sqrt{n}\right\}$. For $k^2\le n<(k+1)^2$ it holds

$$\left\{\sqrt{n}\right\}=\sqrt{n}-k\le\sqrt{(k+1)^2-1}-k,$$

where equality is attained for $n=(k+1)^2-1$. If $k<l$ then

$$1-\left\{\sqrt{(k+1)^2-1}\right\}=k+1-\sqrt{(k+1)^2-1}=$$

$$=\frac{1}{k+1+\sqrt{(k+1)^2-1}}>\frac{1}{l+1+\sqrt{(l+1)^2-1}}=1-\left\{\sqrt{(l+1)^2-1}\right\},$$

V. Brayman and A. Kukush, *Undergraduate Mathematics Competitions (1995–2016)*, Problem Books in Mathematics,
DOI 10.1007/978-3-319-58673-1_31

hence $\left\{\sqrt{(k+1)^2-1}\right\}$ is increasing in k. Thus, for $N = m^2 - 1$ we get

$$\max_{1\le n\le N}\{\sqrt{n}\} = \left\{\sqrt{m^2-1}\right\} = \sqrt{m^2-1} - m + 1.$$

Now consider $m^2 \le N \le (m+1)^2 - 1$. Since $\max_{1\le n\le N}\{\sqrt{n}\}$ is increasing in N, it holds

$$\sqrt{m^2-1} - m + 1 \le \max_{1\le n\le N}\{\sqrt{n}\} \le \sqrt{(m+1)^2-1} - m.$$

Therefore,

$$1 + m - \sqrt{(m+1)^2-1} \le 1 - \max_{1\le n\le N}\{\sqrt{n}\} \le m - \sqrt{m^2-1}.$$

Also we have $m \le \sqrt{N} < m+1$, so

$$\frac{m}{m+1}\cdot(m+1)\left(m+1-\sqrt{(m+1)^2-1}\right) \le$$
$$\le \sqrt{N}\left(1 - \max_{1\le n\le N}\{\sqrt{n}\}\right) < \frac{m+1}{m}\cdot m\left(m - \sqrt{m^2-1}\right).$$

Since $m(m-\sqrt{m^2-1}) = \frac{m}{m+\sqrt{m^2-1}} \to \frac{1}{2}$ and $\frac{m}{m+1} \to 1$, as $m \to \infty$, we have $\lim_{N\to\infty} \sqrt{N}\left(1 - \max_{1\le n\le N}\{\sqrt{n}\}\right) = \frac{1}{2}$.

Answer: $\frac{1}{2}$.

3 For $k < n$ the expansion of $\vec{x}$ in $\vec{x_1}, \dots, \vec{x_k}$ exists not for all $\vec{x}$. For $k = n$ the expansion is unique, and for some $\vec{x}$ it contains negative coefficients. Therefore, $k \ge n+1$. The collection of $k = n+1$ vectors $\vec{x_1} = (1, 0, \dots, 0)$, $\vec{x_2} = (0, 1, 0, \dots, 0)$, $\dots$, $\vec{x_n} = (0, \dots, 0, 1)$, $\vec{x}_{n+1} = (-1, -1, \dots, -1)$ has the required property. Indeed, for every vector $\vec{x}$, one can select $a_{n+1} > 0$ such that all the coordinates of the vector

$$\vec{x} - a_{n+1}\vec{x}_{n+1} = \vec{x} + a_{n+1}(1, 1, \dots, 1)$$

are positive, and then $\vec{x} = \sum_{i=1}^{n} a_i \vec{x_i} + a_{n+1}\vec{x}_{n+1}$, where $a_1, \dots, a_n$ are positive.

Answer: $k = n+1$.

4 For $n = 1$ every matrix is proportional to the identity matrix, and one can take $A = (1)$, $B = (0)$, $\operatorname{rk} A + \operatorname{rk} B = 1 = n$. For $n \ge 2$ let A be the diagonal matrix $A = \operatorname{diag}(0, 1, \dots, n-1)$, $\operatorname{rk} A = n-1$. If a matrix $X = (x_{ij})$ commutes with A then $(i-1)x_{ij} = (j-1)x_{ij}$ for all i, j, whence $x_{ij} = 0$ for $i \ne j$. Thus, X is a diagonal matrix. Let all the entries of the matrix B equal 1. Then $\operatorname{rk} B = 1$, so $\operatorname{rk} A + \operatorname{rk} B = n$. If the matrix $X = \operatorname{diag}(x_{11}, x_{22}, \dots, x_{nn})$ commutes with B, then

for all i, j the equality $x_{ii} = x_{jj}$ holds. Hence the matrix X is proportional to the identity matrix. Therefore, $n \times n$ matrices A and B with required properties exist for every $n \in \mathbb{N}$.
Answer: for all $n \in \mathbb{N}$.

5 By taking the logarithms of both sides of the inequality we get

$$\sum_{k=2}^{n} \frac{1}{k!} \ln k < \ln 2, \ n \geq 2.$$

The function $\ln x$ is concave on $(0, +\infty)$. By Jensen's inequality we have

$$\sum_{k=2}^{n} \frac{1}{k!} \ln k \leq \sum_{k=2}^{n} \frac{1}{k!} \cdot \ln \frac{\sum\limits_{k=2}^{n} \frac{1}{k!} \cdot k}{\sum\limits_{k=2}^{n} \frac{1}{k!}} = \sum_{k=2}^{n} \frac{1}{k!} \ln \left(\frac{\sum\limits_{k=1}^{n-1} \frac{1}{k!}}{\sum\limits_{k=2}^{n} \frac{1}{k!}} \right).$$

Since the sequence $\sum\limits_{k=2}^{n} \frac{1}{k!} \ln k$ is increasing, we get

$$\sum_{k=2}^{n} \frac{1}{k!} \ln k < \sum_{k=2}^{\infty} \frac{1}{k!} \ln k \leq$$

$$\leq \lim_{n\to\infty} \sum_{k=2}^{n} \frac{1}{k!} \ln \left(\frac{\sum\limits_{k=1}^{n-1} \frac{1}{k!}}{\sum\limits_{k=2}^{n} \frac{1}{k!}} \right) = \sum_{k=2}^{\infty} \frac{1}{k!} \ln \left(\frac{\sum\limits_{k=1}^{\infty} \frac{1}{k!}}{\sum\limits_{k=2}^{\infty} \frac{1}{k!}} \right) =$$

$$= (e-2) \ln \frac{e-1}{e-2} = (e-2) \ln \left(1 + \frac{1}{e-2} \right).$$

It remains to show that

$$(e-2) \ln \left(1 + \frac{1}{e-2} \right) < \ln 2, \text{ or } \left(1 + \frac{1}{e-2} \right)^{e-2} < 2.$$

For $a > -1$ the function $y(x) = (1+a)^x$ is convex on $\mathbb{R}$, hence for $x \in (0, 1)$ its graph lies under the secant $y = 1 + ax$, that is $(1+a)^x < 1 + ax$, $0 < x < 1$. In particular for $a = \frac{1}{e-2}$ and $x = e - 2$ we get

$$\left(1 + \frac{1}{e-2} \right)^{e-2} < 1 + \frac{e-2}{e-2} = 2.$$

6 For $n \geq 0$ it holds

$$\frac{x^{3^n} + \left(x^{3^n}\right)^2}{1 - x^{3^{n+1}}} = \frac{1 + x^{3^n} + x^{2\cdot 3^n}}{\left(1 - x^{3^n}\right)\left(1 + x^{3^n} + x^{2\cdot 3^n}\right)} - \frac{1}{1 - x^{3^{n+1}}} = \frac{1}{1 - x^{3^n}} - \frac{1}{1 - x^{3^{n+1}}}.$$

Therefore,

$$\sum_{n=0}^{N} \frac{x^{3^n} + \left(x^{3^n}\right)^2}{1 - x^{3^{n+1}}} = \frac{1}{1 - x} - \frac{1}{1 - x^{3^{N+1}}}.$$

Passing to the limit as $N \to \infty$ we obtain the answer.
Answer:

$$\sum_{n=0}^{\infty} \frac{x^{3^n} + \left(x^{3^n}\right)^2}{1 - x^{3^{n+1}}} = \begin{cases} \frac{1}{1-x}, & |x| > 1, \\ \frac{x}{1-x}, & |x| < 1, \\ 0, & x = -1. \end{cases}$$

7 Consider sets $\mathbb{N}_n = \{1, 2, \ldots, n\}$, $\mathbb{N}_m = \{1, 2, \ldots, m\}$, $m \leq n$, and find the number of surjective mappings from $\mathbb{N}_n$ onto $\mathbb{N}_m$. Let A be the set of all the mappings from $\mathbb{N}_n$ into $\mathbb{N}_m$, and $A_{i_1,\ldots,i_k}$ be the set of all the mappings from $\mathbb{N}_n$ into $\mathbb{N}_m \backslash \{i_1, \ldots, i_k\}$. Then the set of all surjective mappings from $\mathbb{N}_n$ onto $\mathbb{N}_m$ coincides with $A \backslash \bigcup\limits_{i=1}^{m} A_i$. By the inclusion-exclusion principle

$$\left|\bigcup_{i=1}^{m} A_i\right| = \sum_{k=1}^{m} (-1)^{k+1} \sum_{1 \leq i_1 < \ldots < i_k \leq n} |A_{i_1} \cap \ldots \cap A_{i_k}|.$$

A set $A_{i_1} \cap A_{i_2} \cap \ldots \cap A_{i_k} = A_{i_1,\ldots,i_k}$ contains $(m - k)^n$ mappings, because for each element of $\mathbb{N}_n$ there exist $m - k$ possible images. Similarly A contains m^n mappings. Hence

$$\left|\bigcup_{i=1}^{m} A_i\right| = \sum_{k=1}^{m} (-1)^{k+1} \binom{m}{k} (m - k)^n =$$
$$= \sum_{k=0}^{m-1} (-1)^{m-k+1} \binom{m}{m-k} k^n = \sum_{k=0}^{m-1} (-1)^{m+k+1} \binom{m}{k} k^n,$$

$$\left|A \backslash \bigcup_{i=1}^{m} A_i\right| = m^n - \sum_{k=0}^{m-1} (-1)^{k+m+1} \binom{m}{k} k^n = \sum_{k=0}^{m} (-1)^{k+m} \binom{m}{k} k^n.$$

On the other hand, each surjection from $\mathbb{N}_n$ onto $\mathbb{N}_m$ can be constructed as follows: first select m elements that will be "representatives" of inverse images of all distinct elements from $\mathbb{N}_m$, and then define the images of other $n - m$ elements arbitrarily. We get a list of $\binom{n}{m} \cdot m! \cdot m^{n-m}$ mappings which contains all surjections, but some of them might be counted several times, if the choice of "representatives" is not unique. Therefore,

$$\sum_{k=0}^{m} (-1)^{m+k} \binom{m}{k} k^n \leq \binom{n}{m} \cdot m! \cdot m^{n-m},$$

hence

$$\sum_{k=0}^{m} (-1)^{m+k} \binom{m}{k} \left(\frac{k}{m}\right)^n \leq \binom{n}{m} \cdot \frac{m!}{m^m}.$$

8 Denote the vertices of T by A, B, C. Assume that we have to construct a triangle DEF similar to T, such that the vertices D, E, F correspond to A, B, C, respectively. It is sufficient to find the sides of triangle DEF or equivalently, to find the similarity coefficient. Since any two parabolas are similar and the similarity coefficient equals to the ratio of the distances from their centers to the directrices, it suffices to find these distances for the given parabola and the parabola with focus C which passes through the points A and B. Find the directrices of both parabolas. The distances from the points A and B to the directrix of the parabola are equal to the distances from these points to its focus C. Therefore, the directrix is a common tangent to the circles with centers A, B and radii AC, BC. Similarly the directrix of the given parabola is a common tangent to the circles with centers G, H and radii GF, HF, where G and H are arbitrary points on the parabola. Thus, using a compass and a ruler one can find the directrices of the parabolas and distances d_C, d_F from their foci C, F to the directrices. Next find the sides of triangle DEF from the conditions $\frac{DE}{AB} = \frac{EF}{BC} = \frac{DF}{AC} = \frac{d_F}{d_C}$ and construct this triangle.

The problem might have several solutions.

9 Show that for every Lebesgue measurable set $A \subset \mathbb{R}^2$ there exists a Borel measurable set $B \subset A$ such that the set $E = A \setminus B$ has zero measure. If $A \neq \varnothing$ and $\lambda(A) < \infty$, then from the construction of Lebesgue measure it follows that

$$\lambda(A) = \inf \left\{ \sum_{k=1}^{\infty} \lambda(A_k) \,\middle|\, \bigcup_{k=1}^{\infty} A_k \supset A, \{A_k\} \subset K \right\} =$$

$$= \sup \left\{ \sum_{k=1}^{\infty} \lambda(A_k) \,\middle|\, \bigcup_{k=1}^{\infty} A_k \subset A, A_i \cap A_j = \varnothing, i \neq j, \{A_k\} \subset K \right\},$$

where K is the ring generated by sets

$$(x', x''] \times (y', y''], \; -\infty < x' < x'' < \infty, \; -\infty < y' < y'' < \infty.$$

Hence for $\varepsilon > 0$ one can find the sets

$$A_k(\varepsilon) \subset A, \; k \geq 1, \; A_i(\varepsilon) \cap A_j(\varepsilon) = \varnothing, \; i \neq j, \; A_k(\varepsilon) \in K,$$

such that $\bigcup_k A_k(\varepsilon) = A(\varepsilon) \subset A$ is a Borel measurable set and

$$\lambda\left(A(\varepsilon)\right) = \sum_{k=1}^{\infty} \lambda\left(A_k(\varepsilon)\right) > \lambda(A) - \varepsilon,$$

that is $\lambda\left(A \backslash A(\varepsilon)\right) < \varepsilon$. Hence $B = \bigcup_{n \ge 1} A\left(\frac{1}{n}\right)$ is a Borel measurable set, $B \subset A$ and $\lambda(A \backslash B) \le \lambda\left(A \backslash A\left(\frac{1}{n}\right)\right) < \frac{1}{n}$, $n \ge 1$, therefore, $\lambda(A \backslash B) = 0$.

For every Lebesgue measurable set A consider $A^n = A \cap \{x \in \mathbb{R}^n : \|x\| \le n\}$, $n \ge 1$. Since $\lambda(A^n) < \infty$, there exist Borel measurable sets $B^n \subset A^n$ such that $\lambda(A^n \backslash B^n) = 0$. Put $B = \bigcup_{n \ge 1} B^n$. Then $B \subset A$ and $\lambda(A \backslash B) = 0$.

Answer: no, it does not.

10 Notice that $d(X, A)$ is the Hilbert–Schmidt norm of the operator which corresponds to the matrix $X - A$. Hence $d(X, A)$ does not depend on the choice of orthonormal basis. First solve the problem for the most convenient basis, and then come back to the basis from the statement of the problem. Since the matrix A is real and symmetric, there exists an orthonormal eigenbasis $f_1, \ldots, f_n$ such that $Af_k = \lambda_k f_k$, $1 \le k \le n$. Let operators, which correspond to the matrices A and X, are determined by the matrices $(\widetilde{a}_{ij})$ and $(\widetilde{x}_{ij})$ in the basis $\{f_k\}$. Then $\widetilde{a}_{ij} = 0$, $i \ne j$ and $\widetilde{a}_{ii} = \lambda_i$. Hence

$$d^2(X, A) = d^2\left(\widetilde{X}, \widetilde{A}\right) = \sum_{i \ne j} \widetilde{x}_{ij}^2 + \sum_{k=1}^{n} (\widetilde{x}_{kk} - \lambda_k)^2 \ge \sum_{k=1}^{n} \left(\min(\lambda_k, 0)\right)^2,$$

because the positive semidefiniteness of X implies that $\widetilde{x}_{kk} \ge 0$. The equality is attained if and only if $\widetilde{x}_{ij} = 0$, $i \ne j$, and $\widetilde{x}_{kk} = \max(\lambda_k, 0)$. Hence $\{f_k\}$ is a common eigenbasis of the matrices A and X, moreover $Xf_k = \max(\lambda_k, 0) f_k$. Then $Xf = \max(\lambda_k, 0) f$, for each vector f from the subspace corresponding to the eigenvalue λ_k, in particular $Xe_k = \max(\lambda_k, 0)e_k$, $1 \le k \le n$. If S is a matrix with columns $e_1, \ldots, e_n$, then $(\widetilde{a}_{ij}) = S^{-1}AS$ and $(\widetilde{x}_{ij}) = S^{-1}XS$. Therefore, the matrix in question is $X = S(\widetilde{x}_{ij})S^{-1}$.

11 Let ψ be a linear fractional transformation which maps the upper half-plane $\{\mathrm{Im}\, z > 0\}$ onto the disc $\{|\omega| < 1\}$, moreover $\psi(z_0) = 0$. Then ψ maps Ω onto $\{|\omega| < 1\} \backslash \psi(T)$. Put $\Phi = \psi(\varphi^{-1}(\psi^{-1}))$, where φ^{-1} and ψ^{-1} are the inverse transformations to φ and ψ. Then Φ maps $\{|\omega| < 1\}$ onto $\{|\omega| < 1\} \backslash \psi(T)$, and

$$\begin{aligned}\Phi'(0) =& \psi'\left(\varphi^{-1} \circ \psi^{-1}(0)\right) \cdot \left(\varphi'\left(\varphi^{-1} \circ \psi^{-1}(0)\right)\right)^{-1} \cdot \left(\psi'\left(\psi^{-1}(0)\right)\right)^{-1} = \\ =& \psi'(z_0) \cdot \left(\varphi'(z_0)\right)^{-1} \cdot \left(\psi'(z_0)\right)^{-1} = \left(\varphi'(z_0)\right)^{-1}.\end{aligned}$$

Since $|\Phi(\omega)| < 1$ for $|\omega| < 1$ and $\Phi(0) = 0$, by Schwarz lemma it holds $|\Phi(\omega)| \le |\omega|$ for $|\omega| < 1$. Thus, $|\Phi'(0)| \le 1$, whence $|\varphi'(z_0)| = |\Phi'(0)|^{-1} \ge 1$.

12 Let the vertices of the triangle have coordinates

$$A(\cos\alpha, \sin\alpha),\ B(\cos\beta, \sin\beta),\ C(\cos\gamma, \sin\gamma),$$

where α, β, γ are independent identically distributed on $[0, 2\pi)$ random variables. Express the area of the triangle ABC in terms of the areas of triangles AOB, BOC, and AOC, where O is the center of the circle. For all the possible α, β, γ the area of the triangle ABC equals

$$\tfrac{1}{2}\left|\sin(\beta-\alpha)+\sin(\gamma-\beta)+\sin(\alpha-\gamma)\right|.$$

Hence the mean value of the area is

$$S = \tfrac{1}{(2\pi)^3}\int_{[0,2\pi)^3} \tfrac{1}{2}\left|\sin(\beta-\alpha)+\sin(\gamma-\beta)+\sin(\alpha-\gamma)\right| d\alpha\, d\beta\, d\gamma.$$

Notice that for $0 \le \alpha \le \beta \le \gamma < 2\pi$ we have

$$\sin(\beta-\alpha)+\sin(\gamma-\beta)+\sin(\alpha-\gamma) \ge 0,$$

because at least two of the summands are non-negative and their sum is not less than the absolute value of the third one. Hence by symmetry of the integrand in α, β, γ we get $S = \frac{6}{2\cdot(2\pi)^3} I = \frac{3}{8\pi^3} I$, where

$$\begin{aligned}
I &= \iiint\limits_{0\le\alpha\le\beta\le\gamma<2\pi} (\sin(\beta-\alpha)+\sin(\gamma-\beta)+\sin(\alpha-\gamma))\, d\alpha\, d\beta\, d\gamma = \\
&= \int_0^{2\pi}\left(\int_\beta^{2\pi}\left(\int_0^\beta (\sin(\beta-\alpha)+\sin(\gamma-\beta)+\sin(\alpha-\gamma))\, d\alpha\right) d\gamma\right) d\beta = \\
&= \int_0^{2\pi}\int_\beta^{2\pi} (1-\cos\beta+\beta\sin(\gamma-\beta)-\cos(\beta-\gamma)+\cos\gamma)\, d\gamma\, d\beta = \\
&= \int_0^{2\pi} ((2\pi-\beta)(1-\cos\beta)-\beta\cos\beta+\beta)\, d\beta = \\
&= \int_0^{2\pi} 2\pi(1-\cos\beta) d\beta = 4\pi^2.
\end{aligned}$$

Answer: $\frac{3}{2\pi}$.

2004

1 For $n = 1$ the inequality is evident. For $n \geq 2$ it holds

$$\sum_{k=1}^{n} \frac{2k-1}{(k+2)!} = \sum_{k=1}^{n} \left(\frac{2}{(k+1)!} - \frac{5}{(k+2)!} \right) =$$

$$= 2\sum_{k=1}^{n} \frac{1}{(k+1)!} - 5\sum_{k=2}^{n+1} \frac{1}{(k+1)!} =$$

$$= \frac{2}{2!} - 3\sum_{k=2}^{n} \frac{1}{(k+1)!} - \frac{5}{(n+2)!} < \frac{2}{2!} - \frac{3}{3!} = \frac{1}{2}.$$

2 If $n \neq 3k+2$ then $2n-1$ not divides by 3 and covering is impossible. For $n = 3k+2$ covering is possible if and only if the figure can be divided into 2×2 square which contains the erased cell and several 2×3 rectangles. It is possible if the cell was erased from ith column, where i not divides by 3. Hence the probability is $\frac{2k+2}{3k+2}$.
Answer:

$$\begin{cases} 0 & \text{if } n \neq 3k+2, \\ \frac{2k+2}{3k+2} & \text{if } n = 3k+2. \end{cases}$$

3 Change variables in the integrals and get

$$\tfrac{2}{5}\int_0^1 f(x)\,dx + \tfrac{2}{5}\int_0^1 f\left(\tfrac{3}{5}x\right)dx \geq \tfrac{4}{5}\int_0^1 f\left(\tfrac{4}{5}x\right)dx.$$

V. Brayman and A. Kukush, *Undergraduate Mathematics Competitions (1995–2016)*, Problem Books in Mathematics,
DOI 10.1007/978-3-319-58673-1_32

This inequality is obtained by integrating the inequality

$$f(x) + f\left(\tfrac{3}{5}x\right) \geq 2f\left(\tfrac{4}{5}x\right),$$

which follows from the convexity of f.

4 If $f(x_1) = f(x_2)$ then $x_1 = f(f(x_1)) = f(f(x_2)) = x_2$. Thus, the function f is injective. Then the continuity of f implies its monotonicity. Assume that f is increasing. If for some $x_0 \in \mathbb{R}$ it holds $f(x_0) < x_0$ then $f(f(x_0)) < f(x_0) < x_0$, and if $f(x_0) > x_0$ then $f(f(x_0)) > f(x_0) > x_0$, a contradiction. Thus, the only increasing function f is $f(x) \equiv x$. Now assume that f is decreasing. Then $-f$ is increasing and $-f(-f(x)) = f(f(x)) = x$. Therefore, $-f(x) = x$, i.e. $f(x) = -x$, $x \in \mathbb{R}$.
Answer: $f(x) \equiv x$ or $f(x) \equiv -x$.

5 In some coordinate system, equations of the parabola and the circle are $y = x^2$ and $x^2 + (y - R)^2 = R^2$. The circle and the parabola satisfy the condition of the problem if $y = 0$ is a unique nonnegative root of the equation $y + (y - R)^2 = R^2$, whence $R \leq \frac{1}{2}$. Thus, we have to construct a circle with center $(0, \frac{1}{2})$ and radius $\frac{1}{2}$. To construct the coordinate axes draw two parallel chords of the parabola. The straight line l, which passes through their midpoints, is parallel to Oy. Draw a chord orthogonal to l. Its perpendicular bisector is the axis Oy. We find the vertex of the parabola, so we can draw the axis Ox. The bisector of the angle between Ox and Oy intersects the parabola at the point $(1, 1)$. Now it is easy to construct the point $(0, \frac{1}{2})$ and the required circle.

6 It holds $S^3 = ABC \cdot DAB \cdot CDA \cdot BCD = D^{\mathrm{T}}C^{\mathrm{T}}B^{\mathrm{T}}BCD = ABCD = S$.

7 Construct a sequence $\{B_n\}$ of $2^n \times 2^n$ matrices, $n \geq 0$, such that $\sqrt[2^n]{\det B_n} \to \infty$, as $n \to \infty$. Set $B_0 = (1)$ and

$$B_{n+1} = \begin{pmatrix} B_n & B_n \\ -B_n & B_n \end{pmatrix}, \quad n \geq 0.$$

It holds

$$\det B_{n+1} = \det \begin{pmatrix} B_n & B_n \\ -B_n & B_n \end{pmatrix} = \det \begin{pmatrix} B_n & B_n \\ 0 & 2B_n \end{pmatrix} = 2^{2^n} (\det B_n)^2 .$$

Let $\det B_n = 2^{b_n}$. Then $b_0 = 0$ and $b_{n+1} = 2^n + 2b_n$, $n \geq 1$. It is easy to verify by induction that $b_n = n2^{n-1}$, $n \geq 0$. Therefore,

$$\sqrt[2^n]{\det B_n} = 2^{b_n/2^n} = 2^{n/2} \to \infty, \text{ as } n \to \infty.$$

Answer: no, it does not exist.

8 Let $x_k \neq x_m$. Put $y_k = y_m = \sqrt{x_k x_m}$ and $y_n = x_n$ for $n \neq k, n \neq m$. Then for all $n \geq \max\{k, m\}$ it holds

$$x_1 + \ldots + x_n - n\sqrt[n]{x_1 \cdot \ldots \cdot x_n} = (x_k + x_m - 2\sqrt{x_k x_m}) + \\ + (y_1 + \ldots + y_n - n\sqrt[n]{y_1 \cdot \ldots \cdot y_n}) \geq (\sqrt{x_k} - \sqrt{x_m})^2,$$

hence

$$\liminf_{n\to\infty} \left(x_1 + \ldots + x_n - n\sqrt[n]{x_1 \cdot \ldots \cdot x_n}\right) \geq (\sqrt{x_k} - \sqrt{x_m})^2 > 0.$$

Answer: yes, always.

9 Since the permutation maps the identity matrix to itself, there exists a permutation $\{\sigma_1, \sigma_2, \ldots, \sigma_n\}$ of the set $\{1, 2, \ldots, n\}$ such that the cell (k, k) is mapped to the cell (σ_k, σ_k) for all $1 \leq k \leq n$. For $i \neq k$ and $j \neq l$ consider a matrix, which has exactly one nonzero entry in each row and each column, and moreover two of these entries are in the cells (i, j) and (k, l). Since this matrix is transformed into a nonsingular one, cells (i, j) and (k, l) have to be mapped to cells from distinct rows and distinct columns. Hence for every $k \neq i$, $k \neq j$, the image of the cell (i, j) cannot be in the same column or row with the cell (σ_k, σ_k). Thus, the cell (i, j) is mapped to one of the cells (σ_i, σ_j) or (σ_j, σ_i).

Let the cell $(1, 2)$ be mapped to the cell (σ_1, σ_2). Then for each $j \geq 2$ the cell $(1, j)$ is mapped to the cell (σ_1, σ_j), otherwise the cell (σ_1, σ_j) is the image of the cell $(j, 1)$,i.e., the cells $(1, 2)$ and $(j, 1)$ from distinct rows and columns are mapped to cells from the same row, a contradiction. Therefore, for each $j \geq 2$ the cell $(j, 1)$ is mapped to the cell (σ_j, σ_1). Prove that for every $i, j \geq 2$, $i \neq j$, the cell (i, j) is mapped to the cell (σ_i, σ_j). Indeed, if it is mapped to the cell (σ_j, σ_i), then the cells $(j, 1)$ and (i, j) from distinct rows and distinct columns are mapped to the cells from the same row, a contradiction. Thus, the permutation is a composition of the same permutations of rows and columns.

If the cell $(1, 2)$ is mapped to the cell (σ_2, σ_1), we get similarly that the permutation is a composition of the same permutations of rows and columns and of the transposition of matrix. So in both cases the permutation does not change the determinant.

10 For every positive numbers a and b consider

$$f(a, b) = \sup_{u,v\in\mathbb{R}} \int_0^a \int_0^b \cos(2x + u)\cos(2y + v)\, dx\, dy.$$

The properties of least upper bound imply that

$$f(a_0, b_0) \leq \sum_{k=1}^{n} f(a_k, b_k).$$

Since

$$\sup_{u\in\mathbb{R}} \int_0^a \cos(2x+u)dx = \sup_{u\in\mathbb{R}} (\sin a \cos(a+u)) = |\sin a|$$

and similarly

$$\sup_{v\in\mathbb{R}} \int_0^b \cos(2y+v)dy = |\sin b|,$$

it follows that $f(a,b) = |\sin a \sin b|,\ a, b > 0$, whence the required inequality holds.

11 Existence of $\mathsf{E}\xi$ means the convergence of integral

$$\int_{\mathbb{R}} |x|^\alpha \frac{e^{-x^2/2}}{\sqrt{2\pi}} dx = \sqrt{\frac{2}{\pi}} \int_0^\infty x^\alpha e^{-x^2/2} dx =$$
$$= \sqrt{\frac{2}{\pi}} \left(\int_0^1 x^\alpha e^{-x^2/2} dx + \int_1^\infty x^\alpha e^{-x^2/2} dx \right).$$

The first integral converges for $\alpha > -1$, and the second one converges for all $\alpha \in \mathbb{R}$. *Answer:* $\alpha > -1$.

13 The extension exists by the Hahn–Banach Theorem. Let $F_1, F_2 \in X^*$ be two extensions of $f \in G^*$. Then $\frac{1}{2}(F_1 + F_2)$ is an extension of f, and $\left\|\frac{1}{2}(F_1 + F_2)\right\| \geq \|f\|$. But by the triangle inequality $\|\frac{1}{2}(F_1+F_2)\| \leq \frac{1}{2}(\|F_1\|+\|F_2\|) = \|f\|$, whence

$$\|F_1\| = \|F_2\| = \left\|\tfrac{1}{2}(F_1 + F_2)\right\|.$$

Since X^* is strictly normed, it follows that $F_1 = F_2$.

14 It holds

$$R(0) = 0, \quad R(z) = \int_0^z \left(s - 1 + \frac{1}{s+1}\right) ds = \int_0^z \frac{s^2}{1+s} ds,$$

whence

$$|R(ix)| = \left|\int_0^{ix} \frac{s^2}{1+s} ds\right| = \left|\int_0^x \frac{-t^2}{1+it} i\, dt\right| \leq$$
$$\leq \left|\int_0^x \frac{t^2}{|1+it|} dt\right| \leq \left|\int_0^x t^2 dt\right| = \frac{|x|^3}{3}.$$

15 It holds

$$S^3 = ABC \cdot DAB \cdot CDA \cdot BCD = D^{\mathrm{T}} C^{\mathrm{T}} B^{\mathrm{T}} BCD = ABCD = S,$$

hence $S^3 - S = (S+I)(S^2 - S) = 0$, where I is the identity matrix. Because $S = D^{\mathrm{T}}D$ is positive semidefinite, $S+I$ is positive definite and there exists $(S+I)^{-1}$. Therefore, $S^2 - S = (S+I)^{-1}(S^3 - S) = 0$.

16 A couple (x, y) is a stationary point of F for some λ, μ, if maximum or minimum of $\|y - x\|$ is attained at that point under the condition $f_e(x) = f_{-e}(x) = 1$. Geometrically this means that the vector $y - x$ is orthogonal to ellipses $\{f_e(x) = 1\}$ and $\{f_{-e}(x) = 1\}$. Consider basis (e, g) in $\mathbb{R}^2$, where $g \perp e$ and $\|g\| = \|e\|$. Define A such that $Ae = \frac{e}{2\|e\|}$ and $Ag = \frac{g}{3\|g\|}$. Then radius vectors of edges of the minor axis of the ellipse $\{f_e(x) = 1\}$ are $x_1 = e$ and $x_2 = -3e$, and corresponding vectors for the ellipse $\{f_{-e}(x) = 1\}$ are $y_1 = -e$ and $y_2 = 3e$. From that vectors one can form 4 couples (x_i, y_j), $i, j = 1, 2$. Because for that couples $\|x_i - y_j\| \le 6\|e\|$ and the distance between the points of the ellipses $-e + 3g$, $e - 3g$ equals $\sqrt{40}\|e\| > 6\|e\|$, the maximum is attained at some other couple (x_0, y_0) and for the couple which is symmetric to the latter couple with respect to the straight line that contains the vector e. At last, there are two more couples (x, y), for which $x = y$, they are the intersection points of the ellipses.

18 Let be known that the croupier chooses a number from the interval $[a, a+k-1]$. Put $d(k) = \frac{1}{2} + \frac{1}{2k}\mathrm{odd}(k)$, where

$$\mathrm{odd}(k) = \begin{cases} 1 \text{ if } k \text{ is odd,} \\ 0 \text{ if } k \text{ is even.} \end{cases}$$

Notice that the numbers $a, a+1, \dots, a+k-1$ are chosen with prior probability $\frac{1}{k}$. Prove by induction on k that the player, who guesses the number first, will win with probability at least $d(k)$ and his rival will win with probability at least $1 - d(k)$. For $k = 1$, the statement is evident. Suppose that the statement is proved for $k \le n-1$ and show it for $k = n$.

Let the first player call the number a. Then he/she either wins at once (with probability $\frac{1}{n}$), or plays the role of the second player in the situation with $n-1$ numbers. By the induction hypothesis he/she will win with probability at least

$$\frac{1}{n} + \frac{n-1}{n}(1 - d(n-1)) = \frac{1}{2} + \frac{1}{2n}(1 - \mathrm{odd}(n-1)) = d(n).$$

Now, consider chances to win of the second player. If the first one calls some number c out of the interval $[a, a+n-1]$, then the second player plays the role of the first and improves his/her chances. Suppose that the first player calls the number $a+m-1$, $1 \le m \le n$. Then the second one either loses at once (with probability $\frac{1}{n}$), or plays the role of the first in the situation with $m-1$ or $n-m$ numbers and by the induction hypothesis he/she will win with probability at least

$$\frac{m-1}{n}d(m-1)+\frac{n-m}{n}d(n-m)=$$
$$=\frac{n-1}{2n}+\frac{1}{2n}(\mathrm{odd}(m-1)+\mathrm{odd}(n-m))\geq$$
$$\geq\frac{n-1}{2n}+\frac{1}{2n}(1-\mathrm{odd}(n))=1-d(n).$$

By the principle of mathematical induction the statement is proved. The result of the problem is obtained for $k = 2004$ and $a = 1$.

20 If the sequence $\{x_n\}$ exists then for $n \geq 2005$, it holds

$$\|x_1+\ldots+x_n\|^2=\sum_{i=1}^{n}\|x_i\|^2+2\sum_{1\leq i<j\leq n}(x_i,x_j)<n-\frac{n(n-1)}{2004}\leq 0,$$

a contradiction.
Answer: there is no such sequence.

2005

1 *Necessity*. Let $x_n \to a$, as $n \to \infty$. Then

$$\lim_{m\to\infty} |x_n - x_m| = |x_n - a| \text{ and } \lim_{n\to\infty} \limsup_{m\to\infty} |x_n - x_m| = \lim_{n\to\infty} |x_n - a| = 0.$$

Sufficiency. Show that $\{x_n : n \geq 1\}$ is a Cauchy sequence. For every $\varepsilon > 0$, select $N \in \mathbb{N}$ such that

$$\limsup_{m\to\infty} |x_m - x_N| < \frac{\varepsilon}{2},$$

hence

$$\exists M \in \mathbb{N} : \forall m \geq M \quad |x_m - x_N| < \frac{\varepsilon}{2}.$$

From here for all $m_1, m_2 \geq M$, it holds

$$|x_{m_1} - x_{m_2}| \leq |x_{m_1} - x_N| + |x_{m_2} - x_N| < \varepsilon.$$

We proved that a given sequence is a Cauchy one, and by the Cauchy Criterion the sequence converges.
Answer: yes, it is true.

2 Let $A = (a_{ij})$, $B = (b_{ij})$, and $C = (c_{ij})$, $1 \leq i \leq k$, $1 \leq j \leq l$. Then

$$\mathrm{tr}(A(A^{\mathrm{T}} - B^{\mathrm{T}}) + B(B^{\mathrm{T}} - C^{\mathrm{T}}) + C(C^{\mathrm{T}} - A^{\mathrm{T}})) =$$
$$= \sum_{i,j} (a_{ij}^2 - a_{ij}b_{ij} + b_{ij}^2 - b_{ij}c_{ij} + c_{ij}^2 - c_{ij}a_{ij}).$$

V. Brayman and A. Kukush, *Undergraduate Mathematics Competitions (1995–2016)*, Problem Books in Mathematics,
DOI 10.1007/978-3-319-58673-1_33

It remains to notice that for every $a, b, c \in \mathbb{R}$

$$a^2 + b^2 + c^2 - ab - ac - bc = \tfrac{1}{2}\left((a-b)^2 + (b-c)^2 + (c-a)^2\right) \geq 0.$$

3 Let $\tan\alpha = \frac{p}{q}$. Divide each square of the board into rectangles of size $\frac{1}{p} \times \frac{1}{q}$. Then the ball will move along the diagonals of the rectangles only. Assume that the ball will never fall into a hole. Sooner or later it will move along the diagonal of some rectangle at least twice in the same direction, because there exists a finite number of diagonals and directions. But then the movement of the ball will be periodic. Inspecting the movement of the ball, we get that in a period after the start the ball has to return to the corner, from which it was shot, and fall into the hole. We come to a contradiction.

4 It is easy to verify that the function

$$f(x) := \lim_{n\to\infty} \sqrt{1 + \sqrt{x + \sqrt{x^2 + \ldots + \sqrt{x^n}}}}$$

is well-defined for $x \geq 0$ (because the sequence under the limit is nondecreasing and bounded). For $x < y$, it holds

$$f(x) \leq \sqrt{1 + \sqrt{x + \lim_{n\to\infty} \sqrt{y^2 + \ldots + \sqrt{y^n}}}} < f(y),$$

i.e., f is increasing and the equation $f(x) = 2$ has at most one solution.

Show that $f(4) = 2$. Indeed, for $a, b \geq 0$ and $c \geq 1$, we have

$$\left|\sqrt{c+a} - \sqrt{c+b}\right| = \frac{|a-b|}{\sqrt{c+a} + \sqrt{c+b}} \leq \frac{1}{2}\,|a-b|\,.$$

Therefore, the sequences

$$A_n = \sqrt{1 + \sqrt{4 + \ldots + \sqrt{4^{n-1} + \sqrt{4^n}}}}, \quad B_n = \sqrt{1 + \sqrt{4 + \ldots + \sqrt{4^{n-1} + \sqrt{4^n} + 1}}}$$

satisfy $|A_n - B_n| \leq \frac{1}{2^n}$, hence $\lim\limits_{n\to\infty}(A_n - B_n) = 0$. But

$$\sqrt{4^{n-1} + \sqrt{4^n} + 1} = 2^{n-1} + 1 = \sqrt{4^{n-1}} + 1,$$

whence

$$B_n = B_{n-1} = \ldots = B_1 = \sqrt{1 + \sqrt{4} + 1} = 2,$$

therefore, $f(4) = \lim_{n\to\infty} A_n = \lim_{n\to\infty} B_n = 2.$
Answer: $x = 4$.

5 At the plane consider an equilateral triangle with the center at the origin, and let A, B, C be 2×2 matrices which correspond to symmetries of the plane with respect to the altitudes of the triangle. Then the matrices have no common eigenvector, and moreover the matrix $AB = BC = CA$ corresponds to the rotation 120° around the origin.
Answer: yes, there exist.

6 Transform the product of cosines into a sum, using repeatedly the formula

$$\cos\alpha\cos\beta = \tfrac{1}{2}\left(\cos(\alpha+\beta) + \cos(\alpha-\beta)\right).$$

The expression under the integral equals

$$\frac{1}{2^{2003}} \sum \cos(2x \pm 3x \pm 4x \pm \ldots \pm 2005x),$$

where the summation is performed over all 2^{2003} possible choices of signs. The integral $\int_{-\pi}^{\pi} \cos kx\, dx$ vanishes for $k \in \mathbb{Z} \setminus \{0\}$ and equals 2π for $k = 0$. Hence the integral is nonnegative. It remains to notice that there exists at least one summand, for which $k = 2 \pm 3 \pm 4 \pm \ldots \pm 2005 = 0$. It does exist because

$$(2 - 3 - 4 + 5) + (6 - 7 - 8 + 9) + \ldots + (2002 - 2003 - 2004 + 2005) = 0.$$

7 Let $f(x) = x^3 - 3x$, $a_1 = -2$, $a_3 = -1$, $b_1 = 1$, and $b_3 = 2$. Then $f(a_1) = f(b_1)$, $f(a_3) = f(b_3)$, and $f'(c_1) = f'(c_3) = 0$. Take $a_1 < a_2 < a_3$ and $b_1 < b_2 < b_3$ such that $f(b_2) > f(a_2)$, hence it should hold $f'(c_2) > 0$. But $f'(x) = 3(x^2-1)$, and the conditions $f'(c_1) = f'(c_3) = 0$, $c_1 \le c_2 \le c_3$ imply that $f'(c_2) \le 0$. Thus, in this case required c_i, $1 \le i \le 3$, do not exist.
Answer: no, not always.

8 Alongside with a point (x, y, z), the required set S consider points $(\pm x, \pm y, \pm z)$. If at least two of numbers x, y, z are nonzero, then four or eight above mentioned points belong to S. Hence the number of points of S that do not lie on coordinate axes is divisible by 4. Let $k \le R < k + 1$. Then there are $6k + 1$ points on the axes, and hence $2005 \equiv 6k + 1 \pmod 4$, and k should be even. On the other hand, it is not difficult to show that $7 \le R < 8$ and $k = 7$, a contradiction.

9 Let the points A_1, A_2, and A_3 have coordinates $(1, 0)$, $(0, 1)$, and (x, y), respectively. Suppose that a required polynomial P exists, and consider $Q(x, y) =$

$P(1, 0, 0, 1, x, y)$. The triangle $A_1A_2A_3$ is negative if A_3 lies in the first quadrant and positive if A_3 lies in the other quadrants. Hence for each fixed positive y, the polynomial $Q(x, y)$ changes the sign with the vertex A_3 passing over the point $(0, y)$. Then $Q(0, y) = 0$, $y > 0$, and $Q(x, y) = Q_1(x, y) \cdot x$. In a similar way $Q_1(x, y) = Q_2(x, y) \cdot y$, and $Q(x, y) = xy \cdot Q_2(x, y)$. Now $Q_2(x, y)$ changes the sign with passing over the points $(0, y)$, $y < 0$, or $(x, 0)$, $x < 0$, hence

$$Q_2(x, y) = xy \cdot R(x, y), \; Q(x, y) = (xy)^2 R(x, y).$$

For $xy \neq 0$, the polynomial $R(x, y)$ has the same sign as $Q(x, y)$; therefore, by the same reasoning we obtain $R(x, y) = (xy)^2 S(x, y)$, etc., where the degrees of the polynomials $Q, R, S, \ldots$ are decreasing all the time, a contradiction.

10 Show that f is uniformly continuous. By the Arzela–Ascoli theorem the set K is equicontinuous, i.e.,

$$\forall \varepsilon > 0 \; \exists \delta > 0 \; \forall x \in K \; \forall t_1, t_2 \in [0, 1] : |t_1 - t_2| < \delta \Rightarrow |x(t_1) - x(t_1)| < \varepsilon.$$

Let $t_1, t_2 \in [0, 1]$, $|t_1 - t_2| < \delta$, and let x_1, x_2 be the functions at which the values $f(t_1)$, $f(t_2)$ are attained, respectively (such functions exist because the function $g_t(x) = x(t) + x(1 - t)$ is continuous on the compact set K for each $t \in [0, 1]$). Then

$$\begin{aligned} f(t_1) - f(t_2) =& x_1(t_1) + x_1(1 - t_1) - x_2(t_2) - x_2(1 - t_2) \leq \\ \leq& (x_1(t_1) + x_1(1 - t_1)) - (x_2(t_1) + x_2(1 - t_1)) + \\ &+ |x_2(t_1) - x_2(t_2)| + |x_2(1 - t_1) - x_2(1 - t_2)| < 2\varepsilon \end{aligned}$$

(here we use that $g_{t_1}(x_1) \leq g_{t_1}(x_2)$). Similarly, we get the inequality $f(t_2) - f(t_1) < 2\varepsilon$. Hence $|t_1 - t_2| < \delta$ implies $|f(t_1) - f(t_2)| < 2\varepsilon$, and the uniform continuity is proved.

11 In case $\lambda = 0$, a sequence $\{a_n\}$ can be arbitrary.

If $0 < |\lambda| < 1$ then $|a_{n+1}| < |\lambda| \cdot |a_n| + \frac{1}{|\lambda|}$, $n \geq 1$, whence by induction on n we obtain

$$|a_n| < |\lambda|^{n-1}|a_1| + |\lambda|^{n-2} + \ldots + |\lambda| + 1 + \frac{1}{|\lambda|} \leq |a_1| + \frac{1}{|\lambda|(1 - |\lambda|)}.$$

In case $|\lambda| = 1$ put $a_n = \frac{n}{2}\lambda^n$, and in case $|\lambda| > 1$ put $a_n = \lambda^n$. In both cases the sequences are unbounded and satisfy the condition of the problem.
Answer: $0 < |\lambda| < 1$.

12 Let $K \neq 0$ be a supercompact operator. Then for some $x \in X$ it holds $y = Kx \neq 0$, and the image of a bounded set $M = \{\alpha x \mid \alpha \in (0, 1)\}$ is the set $K(M) = \{\alpha y \mid \alpha \in (0, 1)\}$, which is not closed, hence it is not compact. We come to a contradiction.

13 Because $A^2 = AA^{\mathrm{T}} = I$, the matrix A is nonsingular. Hence $A = A^{\mathrm{T}}$, i.e., A is self-adjoint. Then $A = UBU^{\mathrm{T}}$, where U is some orthogonal matrix and B is a diagonal matrix with eigenvalues of the matrix A on the diagonal. Because $A^2 = I$, its eigenvalues satisfy the equation $\lambda^2 = 1$, thus they are equal to ± 1.

14 Let C_n be the circle in $\mathbb{R}^2$ with center $(0, n)$ and radius n,

$$X = \bigcup_{n \geq 1} C_n \setminus (0, 1)^2$$

endowed with Euclidean metrics, and B be the intersection of X and the disc in $\mathbb{R}^2$ with center at the origin and radius 2. Then X is a connected unbounded space, but there is no connected proper subset of X that contains B.
Answer: no, not always.

15 For $\alpha_1 < \alpha_2$, it holds

$$\int_{\mathbb{R}^+} \exp(\alpha_1 (x+1)^t) d\mu(x) < \int_{\mathbb{R}^+} \exp(\alpha_2 (x+1)^t) d\mu(x),$$

and hence it suffices to consider $0 < \alpha < 1$. Let $\alpha < \beta < 1$. Then $\frac{\alpha(x+1)^t}{\beta x^t} \to \frac{\alpha}{\beta} < 1$ as $x \to +\infty$, and hence there exists $c > 0$ such that $\alpha(x+1)^t \leq \beta x^t$ for all $x \geq c$. Thus,

$$\begin{aligned} &\int_{\mathbb{R}^+} \exp(\alpha(x+1)^t)\, d\mu(x) \leq \\ &\quad \leq \int_0^c \exp(\alpha(x+1)^t)\, d\mu(x) + \int_c^\infty \exp(\beta x^t)\, d\mu(x) \leq \\ &\quad \leq \exp(\alpha(c+1)^t) \int_0^\infty \exp(0 \cdot x^t)\, d\mu(x) + \int_0^\infty \exp(\beta x^t)\, d\mu(x) < \infty. \end{aligned}$$

18 It is not difficult to construct a function $f \in C^{(1)}([x_0, x_n])$ such that

$$f(x_i) = y_i,\ i = 0, \ldots, n,$$

and moreover the derivative of f is positive and bounded away from zero at $[x_0, x_n]$, that is $f'(x) \geq \delta$ for some $\delta > 0$ and all $x \in [x_0, x_n]$. For each $\varepsilon > 0$, there exists a polynomial P_ε such that $\max\limits_{x \in [x_0, x_n]} |f'(x) - P_\varepsilon(x)| < \varepsilon$. Set

$$Q_\varepsilon(x) := y_0 + \int_{x_0}^x P_\varepsilon(t)\, dt,\ x \in [x_0, x_n].$$

Then $\max\limits_{x \in [x_0, x_n]} |f(x) - Q_\varepsilon(x)| < \varepsilon(x_n - x_0)$. Let R_ε by a Lagrange polynomial which interpolates the values $y_i - Q_\varepsilon(x_i)$ at points x_i, $i = 0, \ldots, n$. Consider the transformation which maps a collection of numbers $(z_0, \ldots, z_n)$ to the derivative of the Lagrange interpolating polynomial $R' = R'(z_0, \ldots, z_n)$, where $R(x_i) = z_i$, $i = 0, \ldots, n$. This transformation is a linear transformation of finite-dimensional spaces (because $\deg R' \le n$); therefore, it is continuous and there exists $C > 0$ such that

$$\max_{x \in [x_0, x_n]} |R'(x)| \le C \cdot \max_{0 \le k \le n} |z_k|.$$

Choose $\varepsilon > 0$ such that $C(x_n - x_0)\varepsilon + \varepsilon < \delta$. Then for

$$R_\varepsilon = R\left(y_0 - Q_\varepsilon(x_0), \ldots, y_n - Q_\varepsilon(x_n)\right)$$

it holds $\max\limits_{x \in [x_0, x_n]} |R'_\varepsilon(x)| + \varepsilon \le \delta$. Hence the polynomial $Q_\varepsilon + R_\varepsilon$ is required, because

$$\forall x \in [x_0, x_n] \quad Q'_\varepsilon(x) + R'_\varepsilon(x) > P_\varepsilon(x) + \varepsilon - \delta > f'(x) - \delta > 0.$$

2006

1 The condition of the problem is equivalent to each of the following statements: the polynomial $(x^3+x^2+x+1)^n+x^n$ is divisible by $Q(x)=x^4+x^3+x^2+x+1$, or $(-x^4)^n+x^n$ is divisible by $Q(x)$, or $(-1)^n x^{3n}+1$ is divisible by $Q(x)$. We will denote by $\equiv$ the congruence modulo $Q(x)$, that is $P_1(x)\equiv P_2(x)$ if $P_1(x)-P_2(x)$ is divisible by $Q(x)$. It holds $x^5\equiv 1$, $x^6\equiv x$, $x^7\equiv x^2, \dots$, hence if $3n=5a+b$, where $0\le b\le 4$, then $(-1)^n x^{3n}+1\equiv(-1)^n x^b+1$. It remains to notice that for $b\le 4$ the congruence $(-1)^n x^b+1\equiv 0$ holds if and only if we have simultaneously $b=0$ and n is odd, i.e. $n=10k-5$, $k\in\mathbb{N}$.
Answer: $n=10k-5$, $k\in\mathbb{N}$.

2 The condition of the problem implies that the points $(z+1)z^3$, $(z+1)^2z^2$, $(z+1)^3z$, and $(z+1)^4$ are the vertices of an inscribed quadrangle as well. Let z_0 and z_0' be centers of the circles which pass through the first three and the last three of the latter points. Then $z_0'=\frac{z+1}{z}\cdot z_0$. Notice that $z\ne 0$ because all the points are distinct. Since the quadrangle is inscribed, $z_0'=z_0=\frac{z+1}{z}\cdot z_0$, so $z_0=0$. Thus

$$\left|(z+1)z^3\right|=\left|(z+1)^2z^2\right|,$$

hence $|z|=|z+1|$, that is $\operatorname{Re} z=-\frac{1}{2}$.
Answer: $\operatorname{Re} z=-\frac{1}{2}$.

3 Suppose that $(x_i,x_j)<-\frac{1}{n}$ for all $1\le i<j\le n+1$. Then

$$\|x_1+x_2+\ldots+x_{n+1}\|^2=\sum_{i=1}^{n+1}\|x_i\|^2+2\sum_{1\le i<j\le n+1}(x_i,x_j)<n+1-\frac{n(n+1)}{n}=0,$$

a contradiction.

V. Brayman and A. Kukush, *Undergraduate Mathematics Competitions (1995–2016)*, Problem Books in Mathematics,
DOI 10.1007/978-3-319-58673-1_34

We construct an example of $n+1$ unit vectors in $\mathbb{R}^n$ such that $(x_i, x_j) = -\frac{1}{n}$, $1 \le i < j \le n+1$. Use induction on n. For $n = 1$, put $x_1 = 1$, $x_2 = -1$. Assume that the example has been already constructed for $n < k$. Consider $n = k$. Let $e_1, \dots, e_k$ be an orthobasis in $\mathbb{R}^k$. By inductive hypothesis there exist unit vectors $y_1, \dots, y_k$ in $(k-1)$-dimensional space with the basis $\{e_1, \dots, e_{k-1}\}$ such that $(y_i, y_j) = -\frac{1}{k-1}$, $1 \le i < j \le k$. It is easy to verify that the vectors

$$x_i = \frac{\sqrt{k^2-1}}{k} y_i - \frac{1}{k} e_k, \ 1 \le i \le k, \ x_{k+1} = e_k$$

satisfy the required condition, and by the principle of mathematical induction there exists a required example for all $n \ge 1$.
Answer: $-\frac{1}{n}$.

4 Take an arbitrary column vector $x \in \mathbb{R}^m$, $x \neq 0$. Then for the column vector $y = \begin{pmatrix} -Ax \\ x \end{pmatrix} \in \mathbb{R}^{m+n}$ it holds $y \neq 0$, and the positive definiteness implies that

$$y^{\mathrm{T}} \begin{pmatrix} I_m & A \\ A^{\mathrm{T}} & B \end{pmatrix} y = x^{\mathrm{T}}(B - A^{\mathrm{T}}A)x > 0,$$

so the matrix $B - A^{\mathrm{T}}A$ is positive definite.

5 Consider 2×2 matrices $A_x = \begin{pmatrix} 1 & x \\ x & -1 \end{pmatrix}$, $x \in \mathbb{R}$. We have that $A_x^2 = \begin{pmatrix} 1+x^2 & 0 \\ 0 & 1+x^2 \end{pmatrix}$ commutes with every 2×2 matrix, while $A_x A_y = \begin{pmatrix} 1+xy & y-x \\ x-y & 1+xy \end{pmatrix} \neq A_y A_x$ for $x \neq y$.
Answer: yes, there exists.

6 Since the function $f : (0, +\infty) \to \mathbb{R}$ is concave, the sequence $g(n) = f(n+1) - f(n)$ decreases. If $g(n_0) \le 0$ then $g(n) \le 0$ for all $n \ge n_0$, hence the condition $\lim\limits_{x\to+\infty} f(x) = +\infty$ fails. Therefore, $g(n) > 0$, $n \ge 1$, so there exists a finite limit

$$\lim_{n\to+\infty} g(n) = \lim_{n\to+\infty} (f(n+1) - f(n)) = a \ge 0.$$

By Stolz–Cesaro theorem it holds $\lim\limits_{n\to+\infty} \frac{f(n)}{n} = a$, hence $a = 0$. For every $\varepsilon > 0$, there exists $N_0 = N_0(\varepsilon)$ such that for all $n \ge N_0$, it holds $f(n) < f(n+1) < f(n) + \varepsilon$. Take an integer $K > f(N_0)$. Let N_1 be the least number greater than N_0, for which $f(N_1) \ge K$ (there exists such a number because $\lim\limits_{x\to+\infty} f(x) = +\infty$). Then $K - \varepsilon < f(N_1 - 1) < K$ and $\{f(n)\} > 1 - \varepsilon$. Since $\varepsilon > 0$ is arbitrary, $\sup\limits_{n\in\mathbb{N}} \{f(n)\} = 1$.

Remark 1 In a similar way one can prove that the sequence of the fractional parts $\{f(n)\}$, $n \ge 1$, is dense in $[0, 1]$.

7 Consider the function $f(x) = \left((|x| + 3)\ln^2(|x| + 3)\right)^{-1} : \mathbb{R} \to (0, 1)$. This function is continuous. For $n \ge 3$ it holds

$$a_n = \int_{-n}^{n} f(x)\,dx = 2\int_3^{n+3} \frac{dx}{x\ln^2 x} = \frac{2}{\ln 3} - \frac{2}{\ln(n+3)} \to \frac{2}{\ln 3}, \text{ as } n \to \infty,$$

and

$$|b_n| = \int_{-n}^{n} f(x)|\ln f(x)|\,dx = 2\int_3^{n+3} \frac{\ln x + 2\ln\ln x}{x\ln^2 x}\,dx >$$

$$>2\int_3^{n+3} \frac{\ln x}{x\ln^2 x}\,dx = 2\ln\ln(n+3) - 2\ln\ln 3 \to \infty, \text{ as } n \to \infty.$$

Answer: yes, there exists.

8 Introduce new variables $s = 3a + 5b$, $t = 2a + 2b$. There exist infinitely many couples of integers (s, t) such that $P(s-t)+P(s+t) = 0$, because distinct couples (a, b) correspond to distinct couples (s, t). By Taylor's formula it holds

$$P(s-t) + P(s+t) = 2\left(P(s) + \tfrac{1}{2}P''(s)t^2 + \tfrac{1}{4!}P^{(4)}(s)t^4 + \ldots\right) = 0.$$

For s with absolute value large enough, all the coefficients have the same sign and all the powers of t are even, thus, there are no roots. Hence there exists s_0 that corresponds to infinitely many values of t, which implies

$$P(s_0) + \tfrac{1}{2}P''(s_0)t^2 + \tfrac{1}{4!}P^{(4)}(s_0)t^4 + \ldots = 0,$$

in particular $P(s_0) = 0$.

9 For $z > 0$ it holds $\frac{1}{z} = \int_0^1 x^{z-1}dx$, the integral is proper Riemann integral for $z \geq 1$ and improper one for $0 < z < 1$. Then

$$\sum_{i,j=1}^{n} \frac{a_i a_j}{a_i^2 + a_j^2} = \sum_{i,j=1}^{n} a_i a_j \int_0^1 x^{a_i^2+a_j^2-1}dx =$$

$$= \int_0^1 x^{-1} \sum_{i,j=1}^{n} a_i a_j x^{a_i^2+a_j^2} dx = \int_0^1 x^{-1}\left(\sum_{i=1}^{n} a_i x^{a_i^2}\right)^2 dx \geq 0.$$

Remark 1 If the numbers $|a_i|$, $1 \leq i \leq n$, are distinct then the power functions $x^{a_i^2}$, $0 \leq x \leq 1$, $1 \leq i \leq n$, are linear independent, hence the inequality is strict.

Remark 2 For the functions $f_i(t) = t^{a_i^2-1/2}$, $t \in (0, 1)$, from the space $L_2([0, 1])$, it holds

$$(f_i, f_j) = \int_0^1 t^{a_i^2-1/2} \cdot t^{a_j^2-1/2}dt = \frac{1}{a_i^2 + a_j^2}$$

Hence the matrix $B = \left(\frac{1}{a_i^2+a_j^2}\right)_{i,j=1}^{n}$ is positive semidefinite as the Gram matrix of the set of functions $f_1, \ldots, f_n$, hence for all numbers $a_1, a_2, \ldots, a_n \in \mathbb{R} \setminus \{0\}$ and $b_1, b_2, \ldots, b_n \in \mathbb{R}$, a more general inequality $\sum_{i,j=1}^{n} \frac{b_i b_j}{a_i^2 + a_j^2} \geq 0$ holds.

11 See solution of Problem 7.

Remark 1 If f is a probability density function then $\int_{-\infty}^{\infty} f(x) \ln f(x)\,dx$ is called the entropy of f. Problem 11 is about the existence of a probability density function with infinite entropy.

14
(a) It suffices to verify that the distribution of the random variable $\min(\xi, \eta)$ has no atoms, i.e., for each $x \in \mathbb{R}$ it holds $\mathsf{P}(\min(\xi, \eta) = x) = 0$. This is correct because

$$\mathsf{P}(\min(\xi, \eta) = x) \leq \mathsf{P}(\xi = x) + \mathsf{P}(\eta = x) = 0.$$

(b) Similar to (a), for each Borel measurable set B with $\lambda(B) = 0$ (here λ is Lebesgue measure), it holds

$$\mathsf{P}(\min(\xi, \eta) \in B) \leq \mathsf{P}(\xi \in B) + \mathsf{P}(\eta \in B) = 0,$$

because ξ and η have probability density functions. Therefore, the probability distribution of the random variable $\min(\xi, \eta)$ is absolutely continuous with respect to Lebesgue measure, and by Radon–Nikodym theorem the latter random variable has a probability density function.

15 For $\alpha \in (0, 1)$, put $x(\alpha) = (x_1(\alpha), \frac{\alpha}{2}, \frac{\alpha}{3}, \ldots, \frac{\alpha}{n}, \ldots)$ and select a number $x_1(\alpha)$ such that

$$(x_1(\alpha))^2 + \alpha^2 \left(\frac{1}{4} + \frac{1}{9} + \ldots\right) = (x_1(\alpha))^2 + \left(\frac{\pi^2}{6} - 1\right) \alpha^2 = 1.$$

Then $\|x(\alpha)\|_2 = 1$, $\alpha \in (0, 1)$, and for $\alpha \neq \beta$, it holds

$$\sum_{n=1}^{\infty} |x_n(\alpha) - x_n(\beta)| \geq |\alpha - \beta| \sum_{n=2}^{\infty} \frac{1}{n} = \infty.$$

16 The random variable $\dfrac{\xi\eta}{\xi^2 + \eta^2}$ is well-defined because

$$\mathsf{P}\left(\xi^2 + \eta^2 = 0\right) = \mathsf{P}\left(\xi = 0, \eta = 0\right) = \mathsf{P}^2\left(\xi = 0\right) = 0.$$

Random variables ξ and η are independent and identically distributed, therefore, by Fubini's theorem it holds

$$\mathsf{E}\frac{\xi\eta}{\xi^2+\eta^2} = \mathsf{E}\int_0^{+\infty} \xi\eta \exp\left(-t\left(\xi^2+\eta^2\right)\right) dt =$$
$$= \int_0^{+\infty} \mathsf{E}\xi\eta \exp\left(-t\left(\xi^2+\eta^2\right)\right) dt = \int_0^{+\infty} \left(\mathsf{E}\xi \exp\left(-t\xi^2\right)\right)^2 dt \geq 0.$$

The conditions of Fubini's theorem are fulfilled due to the inequality

$$\mathsf{E}\int_0^{+\infty} |\xi\eta| e^{-t(\xi^2+\eta^2)} dt = \mathsf{E}\frac{|\xi\eta|}{\xi^2+\eta^2} \leq \frac{1}{2}.$$

Remark 1 The inequality cannot be replaced with a strict one, because in the case where ξ is symmetrically distributed, it holds $\mathsf{E}\frac{\xi\eta}{\xi^2+\eta^2} = 0$.

17 Consider n-dimensional simplex $K \subset \mathbb{R}^n$ with vertices $v_1, \ldots, v_{n+1}$, that is the convex hull of the vertices

$$K = \operatorname{conv}(v_1, \ldots, v_{n+1}) = \left\{\alpha_1 v_1 + \ldots + \alpha_{n+1} v_{n+1} \,\middle|\, \alpha_i \geq 0,\, \alpha_1 + \ldots + \alpha_{n+1} = 1\right\},$$

where the points $v_1, \ldots, v_{n+1}$ do not lie in a proper subset of $\mathbb{R}^n$. Let $\lambda < n$. For each point $x = \beta_1 v_1 + \ldots + \beta_{n+1} v_{n+1} \in K$, there exists an index j such that $\beta_j \leq \frac{1}{n+1}$. Let the line that passes through the points x and v_j intersects the face of the simplex opposite to v_j at the point $\widetilde{v}_j$. Then

$$\frac{\|\widetilde{v}_j - x\|}{\|v_j - x\|} = \frac{\beta_j}{1-\beta_j} \leq \frac{1}{n} < \frac{1}{\lambda}.$$

Then the image of v_j under the homothety with center x and coefficient $(-\frac{1}{\lambda})$ lies on the extension of the segment with endpoints x and $\widetilde{v}_j$ behind the point $\widetilde{v}_j$, that is the image is out of K. Hence the image of K under the homothety with center x and coefficient $(-\lambda)$ does not contain the point v_j.

Thus, $\lambda \geq n$. For $\lambda = n$, and for each convex compact set $K \subset \mathbb{R}^n$ we will find a point $x \in K$ such that the image of K under the homothety with center x and coefficient $(-n)$ contains K. We may assume that K has positive (n-dimensional) volume, otherwise K lies in a proper subset of $\mathbb{R}^n$ and the problem is reduced to the case of lower dimension. For arbitrary points $v_1, \ldots, v_{n+1}$ in the compact K, consider the volume of the set $\operatorname{conv}(v_1, \ldots, v_{n+1})$. It is a continuous function on the compact K^{n+1}, hence it attains its (positive) maximum at some points $v_1^*, \ldots, v_{n+1}^*$, which are vertices of a n-dimensional simplex.

Show that the point

$$x = \frac{1}{n+1}\sum_{i=1}^{n+1} v_i^*$$

has required property. Under the homothety with center x and coefficient $(-n)$, the hyperplane which contains the points $v_1^*, \dots, v_{i-1}^*, v_{i+1}^*, \dots, v_{n+1}^*$ is mapped to the parallel hyperplane π_i that contains the point v_i^*. There is no point $y \in K$ lying on the other side of π_i compared with $v_1^*, \dots, v_{n+1}^*$, otherwise we can replace v_i^* with y in the collection $v_1^*, \dots, v_{n+1}^*$ and obtain greater volume. Hence K belongs to the simplex which is the intersection of $n+1$ semi-spaces with boundaries π_i, $1 \le i \le n+1$ that contain the points $v_1^*, \dots, v_{n+1}^*$. But this simplex is just an image of K under the latter homothety.
Answer: $\lambda = n$.

18 Denote by I the identity operator in X. Because $\|I - T_n\| \le 2$, $n \ge 1$, it suffices to prove the convergence

$$\lim_{n\to\infty} \|f - T_n f\|_X = 0$$

for a dense in X set of functions. Put

$$(x)_+ = \begin{cases} x, \ x \ge 0; \\ 0, \ x < 0. \end{cases}$$

Since the linear hull of functions $\{1, x, (x-a)_+, a \in (0,1)\}$ contains all the piecewise linear continuous functions on $[0, 1]$, it is dense in X. Hence it suffices to prove that

$$\forall a \in (0,1) \quad \lim_{n\to\infty} \|(x-a)_+ - T_n(x-a)_+\|_X = 0.$$

The operator T_n is nonnegative, therefore, $T_n(x-a)_+ \ge T_n(x-a)$ and $T_n(x-a)_+ \ge T_n 0 = 0$, thus, $T_n(x-a)_+ \ge (T_n(x-a))_+$. Since $(b)_+ - (c)_+ \le (b-c)_+$ for all $b, c \in \mathbb{R}$, we have

$$(x-a)_+ - T_n(x-a)_+ \le (x-a)_+ - (T_n(x-a))_+ \le ((x-a) - T_n(x-a))_+ ,$$

hence $((x-a)_+ - T_n(x-a)_+)_+ \le ((x-a) - T_n(x-a))_+$, and

$$\left\|((x-a)_+ - T_n(x-a)_+)_+\right\|_X \le \left\|((x-a) - T_n(x-a))_+\right\|_X \to 0, \text{ as } n \to \infty.$$

To accomplish the proof we use the following lemma.

Lemma 1 *Let functions $g, h \in X = L_1([0,1])$ satisfy $g(x) \ge 0$, $h(x) \ge 0$, $x \in [0,1]$, and $\|h\|_X \le \|g\|_X$. Then*

$$\|g - h\|_X \le 2\|(g-h)_+\|_X.$$

Proof It is easy to verify under conditions of the lemma, it holds

$$\|(g-h)_+\|_X - \|(h-g)_+\|_X = \|g\|_X - \|h\|_X \geq 0,$$

hence $\|g-h\|_X = \|(g-h)_+\|_X + \|(h-g)_+\|_X \leq 2\|(g-h)_+\|_X$. □

Now, we apply the lemma to $g(x) \equiv (x-a)_+$ and $h = T_n g$ (by the condition of the problem $h(x) \geq 0$, $x \in [0,1]$, and $\|h\|_X = \|T_n g\|_X \leq \|g\|_X$, that is conditions of the lemma are fulfilled). According to the lemma,

$$\|(x-a)_+ - T_n(x-a)_+\|_X \leq 2\left\|((x-a)_+ - T_n(x-a)_+)_+\right\|_X \to 0, \text{ as } n \to \infty,$$

and the statement of the problem is proved.

2007

1 It holds

$$\prod_{k=1}^{n}\frac{(k+p)(k+q)}{(k+r)(k+s)}=\frac{r!s!}{p!q!}\cdot\frac{(n+p)!}{(n+r)!}\cdot\frac{(n+q)!}{(n+s)!}.$$

After canceling out common factors in numerator and denominator of the fraction $\frac{(n+p)!}{(n+r)!}$, it is easy to show that

$$\lim_{n\to\infty}\frac{(n+p)!n^{r-p}}{(n+r)!}=1$$

and in a similar way that $\lim_{n\to\infty}\frac{(n+q)!n^{s-q}}{(n+s)!}=1$. Therefore,

$$\lim_{n\to\infty}\prod_{k=1}^{n}\frac{(k+p)(k+q)}{(k+r)(k+s)}=\frac{r!s!}{p!q!}\cdot\lim_{n\to\infty}n^{p+q-r-s},$$

whence the answer follows.

Answer: $\begin{cases}0 & \text{if } p+q<r+s;\\ \frac{r!s!}{p!q!} & \text{if } p+q=r+s;\\ +\infty & \text{if } p+q>r+s.\end{cases}$

2 Because $k\binom{2n}{k}=k\cdot\frac{(2n)!}{k!(2n-k)!}=2n\cdot\frac{(2n-1)!}{(k-1)!(2n-k)!}=2n\binom{2n-1}{k-1}$, the sum

$$\sum_{k=1}^{n}k\binom{2n}{k}=\sum_{k=1}^{n}2n\binom{2n-1}{k-1}=n\cdot2\sum_{i=0}^{n-1}\binom{2n-1}{i}=n\sum_{i=0}^{2n-1}\binom{2n-1}{i}=n2^{2n-1}$$

is divisible by 8, for all $n\geq 2$.

V. Brayman and A. Kukush, *Undergraduate Mathematics Competitions (1995–2016)*, Problem Books in Mathematics,
DOI 10.1007/978-3-319-58673-1_35

3 We describe a winning strategy for the second player. If the first player places a number a to cell (i, j) of the matrix, then the second player in return places the number $101 - a$ to cell $(11 - i, j)$ (it is clear that he/she can always do that). In the finally formed matrix, the sum of the 1st and 10th rows is equal to the sum of the 2nd and 9th rows, hence the matrix is singular.
Answer: the second player has a winning strategy.

4 Show that the function f is bounded at $[1, +\infty)$. Indeed, for all $x \geq 1$, by Newton–Leibniz formula it holds

$$f(x) = f(1)+\int_1^x f'(t)dt \leq f(1)+\int_1^x \frac{1}{t^4}dt = f(1)+\frac{1}{3}\left(1 - \frac{1}{x^3}\right) < f(1)+\frac{1}{3}.$$

Therefore, if f is unbounded at $(0, +\infty)$, then as a result of the monotonicity $f(x) \to -\infty$, as $x \to 0+$. Further, consider an integer k such that $\frac{\pi}{2} - 2\pi k < f(1)$. By the intermediate value theorem there exist points $0 < s < t < 1$ for which $f(s) = -\frac{\pi}{2} - 2\pi k$ and $f(t) = \frac{\pi}{2} - 2\pi k$. For $x \in [s, t]$, it holds $-\frac{\pi}{2} - 2\pi k \leq f(x) \leq \frac{\pi}{2} - 2\pi k$, $\cos f(x) \geq 0$, hence $f'(x) \leq 1$, $x \in [s, t]$. But by the mean value theorem there exists $\theta \in [s, t]$ such that $f'(\theta) = \frac{f(t)-f(s)}{t-s} = \frac{\pi}{t-s} > \pi > 1$, a contradiction.

5 We will construct a required polynomial of the form

$$P(x) = x^{2007} + cQ(x) + \sum_{i=1}^{1003}(a_i x + b_i)Q_i(x),$$

where $Q(x) = \prod_{j=1}^{1003}(x - j)^2$ and $Q_i(x) = \prod_{\substack{j=1\\ j\neq i}}^{1003}(x - j)^2$, $1 \leq i \leq 1003$. Notice that for each $1 \leq i \leq 1003$, it holds $Q(i) = Q'(i) = 0$ and $Q_j(i) = Q_j'(i) = 0$ if $j \neq i$. Hence

$$\begin{aligned} P(i) &= i^{2007} + (a_i \cdot i + b_i)Q_i(i), \\ P'(i) &= 2007i^{2006} + a_i Q_i(i) + (a_i \cdot i + b_i)Q_i'(i), \end{aligned}$$

and since $Q_i(i) \neq 0$, one can choose constants a_i and b_i in such a way that

$$P(i) = 2i,\ P'(i) = 0,\ i = 1, 2, \ldots, 1003. \qquad (*)$$

Now, one can choose c such that

$$P\left(\tfrac{1}{2}\right) > 2007,\ P\left(\tfrac{3}{2}\right) > 2007,\ \ldots,\ P\left(\tfrac{2005}{2}\right) > 2007. \qquad (**)$$

The conditions (*) and (**) imply that the polynomial P of degree 2007 has exactly 2006 extreme points, namely $x_1 < 1 < x_2 < 2 < \ldots < x_{1003} < 1003$, and the polynomial is monotone at each of the intervals

$$(-\infty, x_1),\ (x_1, 1),\ (1, x_2),\ \ldots,\ (1003, +\infty).$$

Taking into account that $P(i) = 2i$ and $P(x_i) > 2007$, $1 \leq i \leq 1003$, it is easy to verify that the polynomial P has required properties.
Answer: yes, there exists.

6 Let the minute hand makes an angle $\varphi \in [0; 2\pi)$ with the vertical position. At that moment, the hour hand makes one of the angles

$$\psi_k = \tfrac{\varphi+2\pi k}{12} (\mathrm{mod} 2\pi),\ k \in \mathbb{Z},$$

with the vertical position. The hands intersect along a segment of length

$$\max(0, \cos(\psi_k - \varphi)),$$

hence the area in question equals $\frac{1}{2}\int_0^{2\pi} \rho^2(\varphi)d\varphi$, where

$$\rho(\varphi) = \max\left(0, \max_k \cos(\psi_k - \varphi)\right), \varphi \in \mathbb{R}.$$

Notice that

$$\rho(\varphi) = \cos\left(\min_k \left|\tfrac{\varphi+2\pi k}{12} - \varphi\right|\right) = \cos\left(\min_k \left|\tfrac{11\varphi}{12} - \tfrac{\pi k}{6}\right|\right).$$

This function is periodic with period of $\frac{2\pi}{11}$. Since for $\varphi \in \left[-\frac{\pi}{11}, \frac{\pi}{11}\right]$ it holds $\rho(\varphi) = \cos\left(\frac{11\varphi}{12}\right)$, the area of the figure formed by intersections of hands equals

$$\tfrac{1}{2}\int_0^{2\pi} \rho^2(\varphi)d\varphi = \tfrac{11}{2}\int_{-\frac{\pi}{11}}^{\frac{\pi}{11}} \rho^2(\varphi)d\varphi = \tfrac{11}{2}\int_{-\frac{\pi}{11}}^{\frac{\pi}{11}} \cos^2\left(\tfrac{11\varphi}{12}\right)d\varphi =$$

$$= \tfrac{11}{4}\int_{-\frac{\pi}{11}}^{\frac{\pi}{11}} \left(1 + \cos\left(\tfrac{11\varphi}{6}\right)\right)d\varphi = \left(\tfrac{11\varphi}{4} + \tfrac{3}{2}\sin\left(\tfrac{11\varphi}{6}\right)\right)\Big|_{-\frac{\pi}{11}}^{\frac{\pi}{11}} = \tfrac{\pi+3}{2}.$$

Answer: $\frac{\pi+3}{2}$.

7 Look at the behavior of expression $x_1^3 + x_2^3$ when we *move* the variables $x_1 \leq x_2$ *together*, that is, replace them with $x_1 + \varepsilon$ and $x_2 - \varepsilon$, $0 < \varepsilon \leq \frac{x_2 - x_1}{2}$, and when we *move* them *apart*, that is, replace them with $x_1 - \varepsilon$ and $x_2 + \varepsilon$, $\varepsilon > 0$. After moving together, the product x_1x_2 is not decreasing, and after moving apart, it is not increasing. Hence the sum of cubes of two variables

$$x_1^3 + x_2^3 = (x_1 + x_2)((x_1 + x_2)^2 - 3x_1x_2)$$

is not decreasing as two variables move apart, if $x_1 + x_2 \geq 0$, and as they move together, if $x_1 + x_2 \leq 0$.

We will move apart couples of variables x_i, x_j, for which $x_i + x_j \geq 0$, until one of the variables coincides with an end of the segment $[-1; 2]$. That can be done while at least one couple of variables with nonnegative sum remains in the interval $(-1; 2)$. After that we get one of the next two situations:

(1) some variables are equal to 2 or -1, and the rest of the variables have negative pairwise sums (and moreover each of the variables in the interval $(-1; 2)$ has negative sum with -1), or
(2) some variables are equal to 2 or -1, and a single variable in the interval $(-1; 2)$ has nonnegative sum with -1.

In the first case, let x be the mean value of all the variables in $[-1; 2)$. Then $x < 0$ because all pairwise sums are negative. We will move together couples of variables from the interval $[-1; 2)$ as described above, until one of the variables in the couple equals x.

Thus, we can start with an arbitrary collection of points $x_1, \ldots, x_{10}$ and use the described above moving together and apart (which preserve the sum of elements and not decrease the sum of cubes) to reach one of the following collections:

(1) k variables are at a point $x \in [-1; 0]$ and $10 - k$ variables are at the point 2 ($k = 0, \ldots, 10$);
(2) k variables are at the point -1, a single variable equals $x \in [1; 2]$, and $9 - k$ variables are at the point 2 ($k = 0, \ldots, 9$).

From the conditions $x_1 + \ldots + x_{10} = 10$ and $x \in [-1; 0]$ (or $x \in [1; 2]$), we obtain that either $k = 4$ or $k = 5$ for collections of the first type and $k = 3$ for collections of the second type. It remains to examine the following collections:

$$\left(-\tfrac{1}{2}, -\tfrac{1}{2}, -\tfrac{1}{2}, -\tfrac{1}{2}, 2, 2, 2, 2, 2, 2\right), \quad (0, 0, 0, 0, 0, 2, 2, 2, 2, 2),$$
$$(-1, -1, -1, 1, 2, 2, 2, 2, 2, 2).$$

The maximal value of the sum of cubes is equal to 47.5 and attained at the first collection.
Answer: 47.5.

8 Show that a_n equals to the number b_n of permutations of n elements which consist only of the cycles of length 1 and 2. For $n \geq 3$, either the nth element of a permutation is fixed, and then one can determine the permutation of other elements in b_{n-1} ways, or the nth element forms a cycle of length 2 with some of $n - 1$ other elements, and then one can determine the permutation of other elements in b_{n-2} ways. Hence $b_n = b_{n-1} + (n-1)b_{n-2}$, $n \geq 2$, and equalities $a_1 = b_1 = 1$, $a_2 = b_2 = 2$ imply that $a_n = b_n$ for all $n \geq 1$. Fix an arbitrary odd number p and check that $a_p - 1 = b_p - 1$ is divisible by p, or equivalently, the number of permutations with cycles of length 1 and 2 which contain *at least one* cycle of length 2 is divisible by p. For that purpose we show that for each $1 \le k \le \frac{p-1}{2}$, the number of permutations which contain *exactly* k cycles of length 2 is divisible by p. To determine such a permutation one has to choose $2k$ elements which will form the cycles of length 2, and divide them into couples, while an order of couples and of elements in couples does not matter. Therefore, the number of permutations with k cycles of length 2 equals

$$\binom{p}{2k} \cdot \frac{(2k)!}{2^k \cdot k!} = \frac{p}{2^k} \cdot \binom{p-k}{k} \cdot \frac{(p-1)!}{(p-k)!},$$

hence it is divisible by p, because the number p is odd.

9 For each integer k, consider $n \times n$ matrices of the form $A_k = B_k C B_{-k}$, where

$$B_k = \begin{pmatrix} 1 & k & 0 & \dots & 0 & 0 \\ 0 & 1 & 0 & \dots & 0 & 0 \\ 0 & 0 & 1 & \dots & 0 & 0 \\ \dots & \dots & \dots & \dots & \dots & \dots \\ 0 & 0 & 0 & \dots & 1 & 0 \\ 0 & 0 & 0 & \dots & 0 & 1 \end{pmatrix}, \quad C = \begin{pmatrix} 0 & 1 & 0 & \dots & 0 & 0 \\ 0 & 0 & 1 & \dots & 0 & 0 \\ 0 & 0 & 0 & \dots & 0 & 0 \\ \dots & \dots & \dots & \dots & \dots & \dots \\ 0 & 0 & 0 & \dots & 0 & 1 \\ 1 & 0 & 0 & \dots & 0 & 0 \end{pmatrix}$$

(here C is a permutation matrix). Then the matrix A_k has integer entries and since $B_{-k} = B_k^{-1}$ and $C^n = I$, it holds

$$A_k^n = (B_k C B_{-k})^n = \left(B_k C B_k^{-1}\right)^n = B_k C^n B_k^{-1} = B_k B_k^{-1} = I.$$

It remains to notice that $A_k = \begin{pmatrix} k & 1-k^2 \\ 1 & -k \end{pmatrix}$ for $n = 2$, and

$$A_k = \begin{pmatrix} 0 & 1 & k & 0 & \dots & 0 \\ 0 & 0 & 1 & 0 & \dots & 0 \\ 0 & 0 & 0 & 1 & \dots & 0 \\ \dots & \dots & \dots & \dots & \dots & \dots \\ 0 & 0 & 0 & 0 & \dots & 1 \\ 1 & -k & 0 & 0 & \dots & 0 \end{pmatrix}, \quad n \ge 3,$$

that is all the matrices A_k, which correspond to different values of k, are distinct.
Answer: $n \ge 2$.

10 Consider a continuous function $f(x) = x + \ln x$, $x \in (0, 1]$. Since $\lim\limits_{x \to 0+} f(x) = -\infty$ and $f(1) = 1$, by the intermediate value theorem there exists $a \in (0, 1)$ such that $f(a) = 0$. It is clear that $\sin a \ne 0$, $f'(a) = 1 + \frac{1}{a} \ne 0$, hence it holds $\frac{\sin x}{x + \ln x} \sim \frac{\sin a}{(x-a) f'(a)}$ as $x \to a$. Since the integral $\int_a^1 \frac{dx}{x-a}$ is divergent, the initial integral is divergent as well.

Answer: the integral is divergent.

11 Let $B_N = \{x \in \mathbb{R} : f(x) \geq N\}$, $N \geq 0$. Show that there exists M such that $\lambda(B_M) < \infty$. Indeed, otherwise there exists a set $A_0 \subset B_1$, $\lambda(A_0) = 1$, and for $n \geq 1$, one can select subsequently the sets A_n such that $A_n \subset B_{2^n} \setminus \left(\bigcup_{i=0}^{n-1} A_i\right)$ and $\lambda(A_n) = \frac{1}{2^n}$. Put $A = \bigcup_{n=0}^{\infty} A_n$. Then $\lambda(A) = \sum_{n=0}^{\infty} \lambda(A_n) = 2$, but by the choice of the sets A_n it holds

$$\int_A f(x)\,d\lambda(x) = \sum_{n=0}^{\infty} \int_{A_n} f(x)\,d\lambda(x) \geq \sum_{n=0}^{\infty} 2^n \lambda(A_n) = \sum_{n=0}^{\infty} \frac{2^n}{2^n} = +\infty,$$

a contradiction with the condition of the problem.

Show that a constant M, for which $\lambda(B_M) < +\infty$, and the function

$$g(x) = \begin{cases} f(x),\ x \in B_M, \\ 0,\ x \notin B_M \end{cases}$$

have required properties. Indeed, by the construction $f(x) \leq g(x) + M$ for all $x \in \mathbb{R}$, and g is integrable because $\int_{\mathbb{R}} g\,d\lambda = \int_{B_M} f\,d\lambda < +\infty$.

12 Notice that ξ can be expanded as $\xi = m + \sigma\zeta$, where $\zeta \sim N(0; 1)$ is a standard normal random variable. Hence

$$f(\sigma) = \frac{1}{\sqrt{2\pi}} \int_{\mathbb{R}} \frac{e^{-x^2/2}}{1 + e^{m+\sigma x}} dx = \frac{1}{\sqrt{2\pi}} \int_0^{\infty} p(x, \sigma) e^{-x^2/2} dx,$$

where

$$p(x, \sigma) = \frac{1}{1 + e^{m+\sigma x}} + \frac{1}{1 + e^{m-\sigma x}}.$$

Fix an arbitrary $x > 0$ and study the monotonicity of $p(x, \sigma)$ as a function in σ. We have

$$\begin{aligned} p'_\sigma(x, \sigma) &= -\frac{xe^{m+\sigma x}}{(1 + e^{m+\sigma x})^2} + \frac{xe^{m-\sigma x}}{(1 + e^{m-\sigma x})^2} = -\frac{xe^{m+\sigma x}}{(1 + e^{m+\sigma x})^2} + \frac{xe^{m+\sigma x}}{(e^{\sigma x} + e^m)^2} = \\ &= xe^{m+\sigma x} \cdot \frac{-(e^{\sigma x} + e^m)^2 + (1 + e^{m+\sigma x})^2}{(1 + e^{m+\sigma x})^2 (e^{\sigma x} + e^m)^2} = \frac{xe^{m+\sigma x}(e^{2\sigma x} - 1)(e^{2m} - 1)}{(1 + e^{m+\sigma x})^2 (e^{\sigma x} + e^m)^2}. \end{aligned}$$

Hence for all $x > 0$, the sign of $p'_\sigma(x, \sigma)$ coincides with the sign of the expression $e^{2m} - 1$. Therefore,

$$f'(\sigma) = \frac{1}{\sqrt{2\pi}} \int_0^{\infty} p'_\sigma(x, \sigma) e^{-x^2/2} dx \begin{cases} < 0 \text{ if } m < 0; \\ = 0 \text{ if } m = 0; \\ > 0 \text{ if } m > 0. \end{cases}$$

Answer: f decreases if $m < 0$; is a constant if $m = 0$; increases if $m > 0$.

13 Put $C = A + B - I$ and

$$D = A^{-1} + B^{-1} - \tfrac{1}{2}(A^{-1}B^{-1} + B^{-1}A^{-1}) = \tfrac{1}{2}\left(A^{-1}CB^{-1} + B^{-1}CA^{-1}\right).$$

Then for every vector x it holds

$$\begin{aligned}(Dx, x) =& \tfrac{1}{2}\left((A^{-1}CB^{-1}x, x) + (B^{-1}CA^{-1}x, x)\right) = \\ =& \tfrac{1}{2}\left((CB^{-1}x, A^{-1}x) + (A^{-1}x, CB^{-1}x)\right) = (CB^{-1}x, A^{-1}x).\end{aligned}$$

Notice that the matrices $AB^{-1} = AB^{-1/2} \cdot B^{-1/2}$ and $B^{-1/2}AB^{-1/2}$ have common eigenvalues, and moreover the latter matrix is positive definite. Then all the eigenvalues of AB^{-1} are positive. Let $x \neq 0$ be an eigenvector of AB^{-1} that corresponds to an eigenvalue $\lambda > 0$. Then $B^{-1}x = \lambda A^{-1}x$, $(Dx, x) = \lambda(CA^{-1}x, A^{-1}x) > 0$. Therefore, the matrix D cannot be negative definite.
Answer: no, it is impossible.

Remark 1 Let $A = \begin{pmatrix} 1 & 2 \\ 2 & 5 \end{pmatrix}$ as $B = \begin{pmatrix} 5 & 2 \\ 2 & 1 \end{pmatrix}$. Then the matrix $C = \begin{pmatrix} 5 & 4 \\ 4 & 5 \end{pmatrix}$ is positive definite, but the matrix $D = \begin{pmatrix} -3 & 8 \\ 8 & -3 \end{pmatrix}$ has eigenvalues of both signs. Thus, under the conditions of the problem the matrix D is not necessarily positive definite.

14 Let $\deg P = n$. Assume that $\max\limits_{\{|z|=1\}} |P(z)| < 1$. Then for $|z| = 1$, it holds $|P(z)| < |z^n|$, and by Rouche's theorem the polynomials z^n and $z^n - P(z)$ have the same number of roots in the domain $\{z \in \mathbb{C} : |z| < 1\}$ (here the roots are counted according to their multiplicity). But the polynomial z^n has n roots in the latter domain and the polynomial $z^n - P(z)$ has at most $n - 1$ roots there, a contradiction.

2008

1 It suffices to consider $0 < a, b, c < \frac{\pi}{2}$. For $0 < x < \frac{\pi}{2}$, put $f(x) = \ln\left(\frac{\sin x}{x}\right)$ and define by continuity $f(0) = 0$. Because for $0 < x < \frac{\pi}{2}$ it holds

$$f'(x) = \cot x - \frac{1}{x} < 0, \quad f''(x) = -\frac{1}{\sin^2 x} + \frac{1}{x^2} < 0,$$

the function f is decreasing and concave. Hence

$$f(\pi/6) = f(a) + f(b) + f(c) \geq f(0) + f(a+b) + f(c) \geq$$
$$\geq f(0) + f(0) + f(a+b+c) = f(a+b+c),$$

therefore, $a+b+c \geq \frac{\pi}{6}$. As $a, b \to 0+$, it holds $c \to \frac{\pi}{6}$, thus, the required infimum equals $\frac{\pi}{6}$.
Answer: $\frac{\pi}{6}$.

2 Assume that f is not uniformly continuous. Then for some $\varepsilon > 0$, there exist $x_n \in \mathbb{R}$ and $\delta_n > 0$, $n \geq 1$, such that $\delta_n \to 0$ as $n \to 0$ and $|f(x_n+\delta_n) - f(x_n)| \geq \varepsilon$, $n \geq 1$. Fix an arbitrary $N \in \mathbb{N}$ and choose $n \geq 1$ such that

$$\sup_{x\in\mathbb{R}} |f(x-\delta_n) - 2f(x) + f(x+\delta_n)| \leq \frac{\varepsilon}{N}. \qquad (*)$$

Without loss of generality we may assume that $f(x_n + \delta_n) - f(x_n) \geq \varepsilon$. Inequality $(*)$ implies that for all $x \in \mathbb{R}$, it holds

$$f(x + \delta_n) - f(x) \geq f(x) - f(x - \delta_n) - \frac{\varepsilon}{N},$$

V. Brayman and A. Kukush, *Undergraduate Mathematics Competitions (1995–2016)*, Problem Books in Mathematics, DOI 10.1007/978-3-319-58673-1_36

whence by induction on k we get

$$f\left(x_n+(k+1)\delta_n\right)-f\left(x_n+k\delta_n\right)\geq\varepsilon\left(1-\tfrac{k}{N}\right),\quad 0\leq k\leq N-1.$$

Adding the latter inequalities we obtain

$$f(x_n+N\delta_n)-f(x_n)\geq\varepsilon\left(\tfrac{N}{N}+\tfrac{N-1}{N}+\ldots+\tfrac{1}{N}\right)=\tfrac{(N+1)\varepsilon}{2},$$

hence f cannot be bounded, a contradiction.
Answer: yes, it follows.

3 Since f is even, it holds $f'(0)=f'''(0)=0$. Introduce new variable $y=\sqrt{x}$, and use L'Hospital's Rule several times. We get

$$\begin{aligned}g'(x)&=\lim_{x\to 0+}\frac{f(\sqrt{x})-f(0)}{x}=\lim_{y\to 0+}\frac{f(y)-f(0)}{y^2}=\\&=\lim_{y\to 0+}\frac{f'(y)}{2y}=\lim_{y\to 0+}\frac{f'(y)-f'(0)}{2y}=\frac{f''(0)}{2}\end{aligned}$$

and

$$\begin{aligned}g''(x)&=\lim_{x\to 0+}\frac{g'(x)-g'(0)}{x}=\lim_{x\to 0+}\frac{\frac{f'(\sqrt{x})}{2\sqrt{x}}-\frac{f''(0)}{2}}{x}=\\&=\lim_{y\to 0+}\frac{f'(y)-f''(0)y}{2y^3}=\lim_{y\to 0+}\frac{f''(y)-f''(0)}{6y^2}=\\&=\lim_{y\to 0+}\frac{f'''(y)}{12y}=\lim_{y\to 0+}\frac{f'''(y)-f'''(0)}{12y}=\frac{1}{12}f^{(4)}(0).\end{aligned}$$

Answer: $g''(0)=\frac{1}{12}f^{(4)}(0)$.

4 Put $I_k=[2^{2^k},2^{2^{k+1}})$, $k\geq 1$. Then $[4,+\infty)=\bigcup\limits_{k\geq 1}I_k$. If $x\in I_k$ then $x^2\in I_{k+1}$ and $f(x^2)-f(x)=[\log_2\log_2 x]^{-2}=\frac{1}{k^2}$. Hence for $x\in I_k$, from condition (a) we get subsequently

$$\begin{aligned}f(x)=f\left(x^2\right)-\frac{1}{k^2}&=f\left(x^4\right)-\frac{1}{k^2}-\frac{1}{(k+1)^2}=\ldots=\\&=f\left(x^{2^n}\right)-\frac{1}{k^2}-\ldots-\frac{1}{(k+n-1)^2},\ n\geq 1.\end{aligned}\tag{$*$}$$

Let $c=\lim\limits_{t\to+\infty}f(t)$. Then $x^{2^n}\to\infty$, and $f\left(x^{2^n}\right)\to c$ as $n\to\infty$. We pass to the limit in $(*)$ and get

$$f(x) = c - \sum_{m \geq k} \frac{1}{m^2}, \quad x \in I_k, \; k \geq 1. \tag{**}$$

For arbitrary $4 \leq x_1 \leq x_2$, if $x_1 \in I_{k_1}$ and $x_2 \in I_{k_2}$ then $k_1 \leq k_2$. Thus, relation (**) implies that $f(x_1) \leq f(x_2)$, that is, the function f is nondecreasing.

5 Let $x_1, \dots, x_m$ be distinct roots of $P(x)$ with multiplicities $k_1, \dots, k_m$, respectively, $k_1 + \dots + k_m = n$. Polynomials $P(x)$ and $P'(x)$ are divisible by the polynomial $Q(x) = (x - x_1)^{k_1 - 1} \dots (x - x_k)^{k_m - 1}$ of degree $n - m$. Then $R(x) = nP(x) - xP'(x)$ is divisible by $Q(x)$ as well, hence either $R(x)$ is identical zero or $\deg R(x) \geq \deg Q = n - m$. But

$$R(x) = nP(x) - xP'(x) = p_{n-1}x^{n-1} + 2p_{n-2}x^{n-2} + \dots + np_0,$$

therefore, $R(x)$ can be identical zero if and only if $p_{n-1} = p_{n-2} = \dots = p_0 = 0$, i.e., $P(x) \equiv x^n$. Since $P(x)$ has $m \geq 2$ distinct roots, we get a contradiction. Thus, $\deg R \geq n - m$, hence at least one of the coefficients $p_{n-1}, \dots, p_{n-m}$ is nonzero.

6 We use a natural inner product $(u, v) = u^{\mathrm{T}} \cdot \overline{v}$, $u, v \in \mathbb{C}^n$. Denote by L a linear hull of the columns of the matrix A and by $\overline{L}$ a linear hull of vectors which are complex conjugate to the columns of A. Then $\dim L = \dim \overline{L} = \operatorname{rk} A$ and $L \perp \overline{L}$, whence $\operatorname{rk} A \leq \min(\frac{n}{2}, k)$.

It remains to give an example of a matrix with rank equal to $\min([\frac{n}{2}], k)$. Consider

$$A = \begin{pmatrix} 1 & 0 & 0 & \dots & 0 \\ i & 0 & 0 & \dots & 0 \\ 0 & 1 & 0 & \dots & 0 \\ 0 & i & 0 & \dots & 0 \\ \dots & \dots & \dots & \dots & \dots \end{pmatrix},$$

i.e. $A = (a_{jl})$, where

$$a_{jl} = \begin{cases} 1 \text{ if } l = 2j - 1, \; 1 \leq j \leq [\frac{n}{2}]; \\ i \text{ if } l = 2j, \; 1 \leq j \leq [\frac{n}{2}]; \\ 0, \text{ otherwise.} \end{cases}$$

Then $A^{\mathrm{T}} \cdot A = 0$ and $\operatorname{rk} A = \min\left([\frac{n}{2}], k\right)$.
Answer: $\min\left([\frac{n}{2}], k\right)$.

7 Let

$$f(x) = \begin{cases} e^{-1/x^2} \sin\left(e^{1/x^2}\right), & x \neq 0, \\ 0, & x = 0. \end{cases}$$

Then $f \in C(\mathbb{R})$, moreover $f \in C^{(\infty)}(\mathbb{R} \setminus \{0\})$ and

$$|f(x)| \leq e^{-1/x^2} = o(x^n), \text{ as } x \to 0,$$

for each $n \geq 1$. For $x \neq 0$, the derivative equals

$$f'(x) = \frac{2}{x^3} \cdot \left(e^{-1/x^2} \sin\left(e^{1/x^2}\right) - \cos\left(e^{1/x^2}\right)\right).$$

Since f' is unbounded at each neighborhood of zero, it holds that $f \notin C^{(1)}(\mathbb{R})$.
Answer: no, it is not.

8 In case $n \leq 2$, every matrix A satisfies the condition of the problem, because for all matrices it holds $M^S = M^T$.

In case $n \geq 3$, as a result of the linearity it suffices to verify the condition only for matrices B with a single nonzero entry b_{ij}. In particular if $i + j$ is even, then we get that $a_{ki} = 0$ for all $k \neq i$, and $a_{jj} = a_{ii}$. Hence for $n \geq 4$, it holds $A = \operatorname{diag}(\alpha, \beta, \alpha, \beta, \ldots)$, where α and β are some real numbers, and for $n = 3$ the matrix A takes a form $\begin{pmatrix} \alpha & \gamma & 0 \\ 0 & \beta & 0 \\ 0 & \delta & \alpha \end{pmatrix}$. It is easy to verify that for $n \geq 4$ all such matrices A satisfy the condition, and for $n = 3$ only the following matrices do: $A = \begin{pmatrix} \alpha & 0 & 0 \\ 0 & \beta & 0 \\ 0 & 0 & \alpha \end{pmatrix}$.
Answer: A is arbitrary for $n \leq 2$; $A = \operatorname{diag}(\alpha, \beta, \alpha, \beta, \ldots)$, $\alpha, \beta \in \mathbb{R}$ for $n \geq 3$.

9 For an unmarked point (x, y), define $R(x, y)$ in the same manner as for marked ones, i.e., $R(x, y)$ is the number of marked points (x', y') with $x' \geq x$ and $y' \geq y$. Then for each marked point, it holds

$$R(x, y) - R(x, y + 1) - R(x + 1, y) + R(x + 1, y + 1) = 1.$$

Indeed, on the left-hand side all the points except (x, y) are counted with plus and minus sign the same number of times, and the point (x, y) is counted only in $R(x, y)$. Hence if the point (x, y) is marked then among the numbers $R(x, y)$, $R(x, y + 1)$, $R(x + 1, y)$, and $R(x + 1, y + 1)$ there exists at least one odd number, moreover that number corresponds to a marked point, because for an unmarked point (a, b), it is evident that $R(a, b) = 0$.

Divide the set of points with positive integer coordinates into subsets of the form

$$\{(2r - 1, 2s - 1), (2r - 1, 2s), (2r, 2s - 1), (2r, 2s)\}, \ (r, s) \in \mathbb{N},$$

and consider those of them, which contain at least one marked point. We will get at least $n/4$ such subsets and by the condition of the problem in each such subset the point $(2r - 1, 2s - 1)$ is marked. Thus, each such subset contains a marked point (x, y) with odd value of $R(x, y)$. This proves the required statement.

11 For arbitrary number $a \geq 0$, as a result of independence of ξ and ξ^2 it holds

$$\begin{aligned}\mathsf{P}(|\xi| < a) &= \mathsf{P}\left(|\xi| < a, \xi^2 < a^2\right) = \\ &= \mathsf{P}(|\xi| < a)\,\mathsf{P}\left(\xi^2 < a^2\right) = \mathsf{P}^2(|\xi| < a),\end{aligned}$$

whence either $\mathsf{P}(|\xi| < a) = 0$ or $\mathsf{P}(|\xi| < a) = 1$. Thus, the cumulative distribution function of random variable $|\xi|$ has a single jump point a_0, hence $|\xi| = a_0$, almost surely, and $\cos\xi = \cos|\xi| = \cos a_0$, almost surely.

13 Since f is even, it holds $f'(0) = f'''(0) = \ldots = f^{(2k-1)}(0) = 0$. Introduce the function

$$F(x) = f(x) - \left(f(0) + \frac{1}{2}f''(0)x^2 + \ldots + \frac{1}{(2k)!}f^{(2k)}(0)x^{2k}\right), \quad x \in \mathbb{R}.$$

Then F is $2k$ times differentiable and even function on $\mathbb{R}$, moreover $F^{(n)}(0) = 0$, $0 \leq n \leq 2k$. Therefore,

$$g(x) = f(0) + \frac{1}{2}f''(0)x + \ldots + \frac{1}{(2k)!}f^{(2k)}(0)x^k + G(x),$$

where $G(x) = F(\sqrt{x})$, $x \geq 0$.

Show that the function G is k times differentiable at $x = 0$ and $G^{(k)}(0) = 0$. This will imply that g will be k times differentiable as well, and moreover

$$g^{(k)}(0) = \frac{k!}{(2k)!}f^{(2k)}(0).$$

By Taylor's formula with Peano's form of remainder for each $0 \leq j \leq 2k-1$, it holds $F^{(j)}(x) = o\left(x^{2k-j}\right)$, as $x \to 0+$. Therefore,

$$F^{(j)}(\sqrt{x}) = o\left(x^{k-\frac{j}{2}}\right), \quad \text{as } x \to 0+, \quad 0 \leq j \leq k \leq 2k-1.$$

For $x > 0$ and $1 \leq i \leq k$, we get

$$G^{(i)}(x) = \sum_{1 \leq j \leq i} c_{ji}F^{(j)}(\sqrt{x})x^{-i+\frac{j}{2}} = \sum_{1 \leq j \leq i} c_{ji} \cdot o(x^{k-i}), \text{ as } x \to 0+,$$

where c_{ji} are some real numbers. Hence for $0 \leq i \leq k-1$, it holds $G^{(i)}(x) = o(x)$, as $x \to 0+$.

This makes it possible to prove subsequently for $i = 1, \dots, k$, that there exists $G^{(i)}(0) = 0$. Indeed, assume that there exists $G^{(i-1)}(0) = 0$. Then

$$G^{(i)}(0) = \lim_{x \to 0+} \frac{G^{(i-1)}(x) - G^{(i-1)}(0)}{x} = \lim_{x \to 0+} \frac{o(x) - 0}{x} = 0.$$

Answer: $g^{(k)}(0) = \frac{k!}{(2k)!} f^{(2k)}(0)$.

14 Denote $\Delta = \max\limits_{1 \le i, j \le n} |a_{ij} - b_{ij}|$. The real matrices A and B are symmetric, hence we have the following bound (here x is a column vector and $\|x\|$ is Euclidean norm):

$$\begin{aligned} |\lambda_{\min}(A) - \lambda_{\min}(B)| &= \left| \min_{\|x\|=1} x^{\mathrm{T}} A x - \min_{\|x\|=1} x^{\mathrm{T}} B x \right| = \\ &= \left| \max_{\|x\|=1} x^{\mathrm{T}} (-A) x - \max_{\|x\|=1} x^{\mathrm{T}} (-B) x \right| \le \end{aligned}$$

$$\begin{aligned} &\le \max_{\|x\|=1} \left| x^{\mathrm{T}}(-A)x - x^{\mathrm{T}}(-B)x \right| = \max_{\|x\|=1} \left| x^{\mathrm{T}}(B - A)x \right| \le \\ &\le \max_{\|x\|=1} \left| \sum_{i,j=1}^{n} (b_{ij} - a_{ij}) x_i x_j \right| \le \Delta \cdot \max_{\|x\|=1} \sum_{i,j=1}^{n} |x_i x_j| = \end{aligned}$$

$$= \Delta \cdot \max_{\|x\|=1} \left(\sum_{i=1}^{n} |x_i| \right)^2 \le \Delta \cdot \max_{\|x\|=1} \left(n \cdot \sum_{i=1}^{n} x_i^2 \right) = n\Delta.$$

15 Show that for all $x \in \mathbb{R}^2$, the matrix $J(x) = Df(x) + Df(x)^{\mathrm{T}}$ is positive definite. Indeed, $J(x)$ is a symmetric matrix and according to Problem 14 the function $\lambda(x) = \lambda_{\min}(J(x))$ is continuous on $\mathbb{R}^2$. But by the condition of the problem $\lambda(0) = 2 > 0$ and $\lambda(x) \neq 0$, $x \in \mathbb{R}^2$. Hence for each $x \in \mathbb{R}^2$, it holds $\lambda(x) > 0$, that is, $J(x)$ is positive definite.

Suppose that $f(y) = f(z)$, for some $y \neq z$. For the function

$$g(t) = (z - y)^{\mathrm{T}} f(y + t(z - y)), \ t \in [0, 1],$$

it holds $g(0) = g(1)$. By Rolle's theorem there exists $\theta \in (0, 1)$ such that $g'(\theta) = 0$. Denote $x_\theta = y + \theta(z - y)$. Then

$$\begin{aligned} g'(\theta) = (z - y)^{\mathrm{T}} Df(x_\theta)(z - y) &= \\ &= \tfrac{1}{2}(z - y)^{\mathrm{T}} \left(Df(x_\theta) + Df(x_\theta)^{\mathrm{T}} \right) (z - y) = \\ &= \tfrac{1}{2}(z - y)^{\mathrm{T}} \left(J(x_\theta) \right) (z - y) = 0, \end{aligned}$$

and because of $z - y \neq 0$, we come to a contradiction with the fact that the matrix $J(x_\theta)$ is positive definite. Thus, f is an injection.
Answer: yes, it is.

16 Let F be the cumulative distribution function of ξ. Then F is a continuous bijection of $\mathbb{R}$ onto $(0, 1)$, moreover the random variable $F(\xi)$ is uniformly distributed on $(0, 1)$. But then the random variable $1 - F(\xi)$ is uniformly distributed on $(0, 1)$ as well and one can take $f(x) = F(x)$ and $g(x) = 1 - f(x)$, $x \in \mathbb{R}$. It remains to notice that f does not coincide with g because f is increasing and g is decreasing.
Answer: yes, it is true.

17 Prove, e.g., that $\lambda(\{x : f(x) \geq 1\}) = 0$. Denote $B = \{x : f(x) \geq 1\}$. Suppose that $\lambda(B) = \varepsilon > 0$. Since

$$\int_{[0,1]} f^{2n-1}(x)\, d\lambda(x) = \int_{(0,1)} f^{2n-1}(x)\, d\lambda(x) \to 0, \text{ as } n \to \infty,$$

for every closed set $F \subset [0, 1]$ it holds

$$\int_F f^{2n-1}(x)\, d\lambda(x) = \int_{[0,1]} f^{2n-1}(x)\, d\lambda(x) - \int_{[0,1]\setminus F} f^{2n-1}(x)\, d\lambda(x) \to 0, \text{ as } n \to \infty.$$

The regularity of Lebesgue measure on $[0, 1]$ implies that there exists a closed set $F \subset B$ such that $\lambda(F) \geq \frac{\varepsilon}{2}$. Then

$$\int_F f^{2n-1}(x)\, d\lambda(x) \geq \int_F 1\, d\lambda(x) \geq \frac{\varepsilon}{2} \not\to 0, \text{ as } n \to \infty,$$

a contradiction. In a similar way $\lambda(\{x : f(x) \leq -1\}) = 0$.

18 Notice that M is a group with respect to the multiplication of matrices. Find the order of the group. The first column of a nonsingular 3×3 matrix over $\mathbb{Z}_2$ can be chosen in $2^3 - 1 = 7$ ways (it can be any nonzero vector), the second one in $2^3 - 2 = 6$ ways (it can be any vector that is linearly independent with the first one), and the third one in $2^3 - 2^2 = 4$ ways (it can be any vector which is linearly independent with the first two vectors). Hence M consists of $7 \cdot 6 \cdot 4 = 168 = 8 \cdot 3 \cdot 7$ matrices. By Sylow's theorem M has elements of order 3 and 7. Moreover since $\begin{pmatrix} 1&0&1\\ 0&1&1\\ 1&1&1 \end{pmatrix}^4 = I$ but $\begin{pmatrix} 1&0&1\\ 0&1&1\\ 1&1&1 \end{pmatrix}^2 \neq I$, the group M has an element of order 4. Thus, the least number n such that $A^n = I$ for all $A \in M$, is a divisor of $8 \cdot 3 \cdot 7$ and is divisible by $4 \cdot 3 \cdot 7$.

Show that $n = 4 \cdot 3 \cdot 7 = 84$. It suffices to prove that M has no element of order 8, that is, $A^8 = I$ and $A \in M$ imply that $A^4 = I$. Over $\mathbb{Z}_2$, it holds

$$A^8 - I = (A^4 - I)(A^4 + I) = (A^4 - I)^2 = \ldots = (A - I)^8 = 0.$$

Hence the polynomial $p(t) = (t-1)^8$ annihilates the matrix A. But the minimal polynomial of A is a divisor of $p(t)$ of at most third degree. Therefore, the minimal polynomial of A is a divisor of the polynomial $(t-1)^3$. Hence $(t-1)^4$ annihilates the matrix A, that is $(A-I)^4 = A^4 - I = 0$. Thus, $n = 4 \cdot 3 \cdot 7 = 84$.
Answer: $n = 84$.

2009

1 A quadrilateral $ABCD$ is circumscribed if and only if $AD + BC = AB + CD$. Let R be the radius of the circle circumscribed about the quadrilateral $ABCD$, $\angle ABC = \beta$ and $\angle ABD = \varphi \leq \beta$. Then $\angle DBC = \beta - \varphi$, and by the sine law

$$AD + BC - AB - CD = BC - AB + 2R(\sin\varphi - \sin(\beta - \varphi)) =: f(\varphi).$$

The function f is continuous on $[0, \beta]$ and by the triangle inequality $f(0) = BC - AB - AC < 0$ and $f(\beta) = BC - AB + AC > 0$. By the intermediate value theorem there exists φ_0 in $0 < \varphi_0 < \beta$ such that $f(\varphi_0) = 0$, and the corresponding quadrilateral $ABCD$ is circumscribed.

Alternative solution. In the case $AB \neq BC$ consider a branch of the hyperbola with foci A and C that passes through the point B. This branch intersects the circumcircle of the triangle ABC for the second time at a point D such that $AD - CD = AB - BC$, that is, $AD + BC = AB + CD$, and the quadrangle $ABCD$ is circumscribed.

In the case $AB = BC$ the bisector of AC intersects the circumcircle of the triangle ABC in the required point D.
Answer: yes, there exists.

2 For $n = 1$, the polynomial x^2+x-1 has irrational roots $\frac{-1\pm\sqrt{5}}{2}$, and is irreducible in $\mathbb{Q}[x]$. For $n \geq 2$, it holds

$$F_n x^{n+1} + F_{n+1}x^n - 1 = (x^2 + x - 1)\left(F_n x^{n-1} + F_{n-1}x^{n-2} + \ldots + F_2 x + F_1\right),$$

thus the polynomial is not irreducible.
Answer: $n = 1$.

V. Brayman and A. Kukush, *Undergraduate Mathematics Competitions (1995–2016)*, Problem Books in Mathematics,
DOI 10.1007/978-3-319-58673-1_37

3 (a) The inequality is just an identity. Indeed, it holds

$$\frac{\cos A}{\sin B\sin C}+\frac{\cos B}{\sin A\sin C}+\frac{\cos C}{\sin A\sin B}=\frac{\cos A\sin A+\cos B\sin B+\cos C\sin C}{\sin A\sin B\sin C}=\frac{\sin 2A+\sin 2B+\sin 2C}{2\sin A\sin B\sin C}=$$
$$=\frac{2\sin(A+B)\cos(A-B)-2\sin(A+B)\cos(A+B)}{2\sin A\sin B\sin(A+B)}=\frac{2\sin A\sin B}{\sin A\sin B}=2.$$

(b) By the Cauchy–Schwarz inequality

$$\left(\frac{\cos A}{\sqrt{\sin B\sin C}}+\frac{\cos B}{\sqrt{\sin A\sin C}}+\frac{\cos C}{\sqrt{\sin A\sin C}}\right)^2\le$$
$$\le\left(\frac{\cos A}{\sin B\sin C}+\frac{\cos B}{\sin A\sin C}+\frac{\cos C}{\sin A\sin C}\right)(\cos A+\cos B+\cos C)=$$
$$=2\,(\cos A+\cos B+\cos C)\le\sqrt{3},$$

because the cosine function is concave on $[0,\frac{\pi}{2}]$, and by Jensen's inequality it holds $\cos A+\cos B+\cos C\le 3\cos\frac{A+B+C}{3}=3\cos\frac{\pi}{3}=\frac{\sqrt{3}}{2}$.

4 The trace of a product of two matrices does not depend on the order of multipliers, therefore,

$$\operatorname{tr}ABC=\operatorname{tr}(AB)C=\operatorname{tr}C(AB)=\operatorname{tr}CAB=\operatorname{tr}(CA)B=\operatorname{tr}B(CA)=\operatorname{tr}BCA,$$

and $n=\operatorname{tr}I=\operatorname{tr}(ABC+BCA+CAB)=3\operatorname{tr}ABC$. Thus, n has to be divisible by 3. For $n=3$, one can take

$$A_1=\begin{pmatrix}0&1&0\\0&0&0\\0&0&0\end{pmatrix},\quad B_1=\begin{pmatrix}0&0&0\\0&0&1\\0&0&0\end{pmatrix},\quad C_1=\begin{pmatrix}0&0&0\\0&0&0\\1&0&0\end{pmatrix}.$$

For $n=3k$, $k>1$, one can construct A,B,C as block matrices: $A_k=\operatorname{diag}(A_1,\dots,A_1)$, $B_k=\operatorname{diag}(B_1,\dots,B_1)$, $C_k=\operatorname{diag}(C_1,\dots,C_1)$.
Answer: $n=3k$, $k\in\mathbb{N}$.

5 Denote $K=\{(x(t),y(t))\,|\,t\in\mathbb{R}\}\subset\mathbb{R}^2$. If $(x_1,y_1),(x_2,y_2)\in K$ and $x_1+y_1\le x_2+y_2$, then $x_1\le x_2$ and $y_1\le y_2$, moreover $x_1+y_1=x_2+y_2$ implies that $x_1=x_2$, $y_1=y_2$. Hence on the set $L=\{x+y\,|\,(x,y)\in K\}\subset\mathbb{R}$ the following nondecreasing functions are well-defined: $f:x+y\mapsto x$ and $g:x+y\mapsto y$, where $(x,y)\in K$. Put $z(t)=x(t)+y(t)$, $t\in\mathbb{R}$. Then $x(t)=f(z(t))$, $y(t)=g(z(t))$, $t\in\mathbb{R}$.

It remains to extend f and g from L to $\mathbb{R}$ in such a way that the extensions will be nondecreasing. Show that $f,g:L\to\mathbb{R}$ are Lipschitz continuous with the Lipschitz constant equal to 1. Indeed, for $z_1=x_1+y_1<z_2=x_2+y_2$ it holds

$$|z_2-z_1|=x_2+y_2-x_1-y_1=$$
$$=|x_2-x_1|+|y_2-y_1|=|f(z_2)-f(z_1)|+|g(z_2)-g(z_1)|.$$

Hence f and g can be extended first by continuity to the closure $\overline{L}$, and then in a linear way to each of the intervals that compose $\mathbb{R}\backslash\overline{L}$ preserving the monotonicity of f and g.

6 Denote $y_n = \frac{1}{n}\sum\limits_{k=1}^{n} x_k$, $n \geq 1$. Then $x_k = ky_k - (k-1)y_{k-1}$, $k \geq 1$, where $y_0 = 0$, and

$$\frac{1}{n^p}\sum_{k=1}^{n} k^{p-1}x_k = -\frac{1}{n^p}\sum_{k=1}^{n-1}\left((k+1)^{p-1} - k^{p-1}\right)ky_k + y_n.$$

Put $c_{nk} = \frac{p}{(p-1)n^p}\left((k+1)^{p-1} - k^{p-1}\right)k$, $1 \leq k \leq n-1$, $n \geq 1$. Then $c_{nk} \geq 0$, for each k it holds $c_{nk} \to 0$, $n \to \infty$, and by Stolz-Cesaro theorem

$$\lim_{n\to\infty}\sum_{k=1}^{n-1} c_{nk} = \lim_{n\to\infty}\frac{p}{(p-1)n^p}\sum_{k=1}^{n-1}\left((k+1)^{p-1} - k^{p-1}\right)k =$$

$$= \lim_{n\to\infty}\frac{p}{p-1}\cdot\frac{\left((n+1)^{p-1} - n^{p-1}\right)n}{(n+1)^p - n^p} = 1.$$

This implies by Toeplitz's theorem that $\lim\limits_{n\to\infty}\sum\limits_{k=1}^{n-1} c_{nk}y_k = \lim\limits_{n\to\infty} y_n$, hence

$$\lim_{n\to\infty}\frac{1}{n^p}\sum_{k=1}^{n} k^{p-1}x_k = \lim_{n\to\infty}\left(-\frac{p-1}{p}\sum_{k=1}^{n} c_{nk}y_k + y_n\right) = \frac{1}{p}\lim_{n\to\infty} y_n.$$

7 The function $K(t) = te^{-t}$, $t \geq 0$ increases on $[0; 1]$ and decreases on $[1; +\infty)$. Define $a = a(n)$ by the condition $K(a) = K\left((n-1)a\right)$, that is

$$ae^{-a} = (n-1)ae^{-(n-1)a}, \quad a = \frac{1}{n-2}\ln(n-1).$$

Show that the supremum in question equals $K(a) = \dfrac{\ln(n-1)}{(n-2)(n-1)^{\frac{1}{n-2}}}$ and is attained at points $x_1^* < x_2^* < \ldots < x_n^*$ under the condition

$$x_2^* - x_1^* = x_3^* - x_2^* = \ldots = x_n^* - x_{n-1}^* = a.$$

Indeed, for all $i \neq j$ it holds $|x_i^* - x_j^*| = |i-j|a \in [a; (n-1)a]$ and $K(|x_i^* - x_j^*|) \geq K(a)$, whence $\inf\limits_{i\neq j} K(|x_i^* - x_j^*|) = K(a)$. Assume that $\inf\limits_{i\neq j} K(|x_i - x_j|) > K(a)$ for some points $x_1 < x_2 < \ldots < x_n$. Then it holds $x_2 - x_1 > a$, $x_3 - x_2 > a$, ..., $x_n - x_{n-1} > a$, whence $x_n - x_1 > (n-1)a$. But then $K(|x_n - x_1|) < K\left((n-1)a\right) = K(a)$, a contradiction.

Answer: $\dfrac{\ln(n-1)}{(n-2)(n-1)^{\frac{1}{n-2}}}$.

8 Put $f\left(\frac{p}{q}\right) = \frac{1}{q}$, where $p \in \mathbb{Z}$, $q \in \mathbb{N}$, and $\frac{p}{q}$ is a fraction in lowest terms. Show that the function f has the required properties. Indeed, for $x = \frac{p}{q}$ and $y = \frac{r}{s}$, and $x \neq y$, it holds

$$|x - y| = \frac{|ps - qr|}{qs} \geq \frac{1}{qs} = f(x)f(y).$$

Now, let $x = \frac{p}{q}$ be a fraction in lowest terms. From the Euclidean Algorithm it follows that there exist numbers $r \in \mathbb{Z}$ and $s \in \mathbb{N}$ such that $|ps - qr| = 1$. Then

$$|p(s + nq) - q(r + np)| = 1,$$

and for each $y_n = \frac{r+np}{s+nq}$, $n \geq 1$, it holds $f(x)f(y_n) = |x - y_n|$.
Answer: yes, there exists.

9 For $n = 2$, it is obvious that there exists a required enumeration. For $n = 3$, adjacent indices can be attributed only to permutations that can be obtained from each other by a cyclic shift, hence it is easy to check that there is no required enumeration.

Construct an example for $n \geq 4$. We call an enumeration $\sigma_1, \dots, \sigma_{n!}$ of permutations of the set $\{1, \dots, n\}$:

(A)-enumeration if for all $1 \leq k \leq n$ it holds

$$\sigma_{l+1}(k) \neq \sigma_l(k),\ \ 1 \leq l \leq n! - 1,\ \text{ and } \sigma_1(k) \neq \sigma_{n!}(k);$$

(B)-enumeration if for all $1 \leq k \leq n - 1$ it holds

$$\sigma_{l+1}(k+1) \neq \sigma_l(k),\ \ 1 \leq l \leq n! - 1,\ \text{ and } \sigma_1(k+1) \neq \sigma_{n!}(k).$$

We will show that there exists a required enumeration, i.e., an (A)-enumeration of permutations of the set $\{1, \dots, n\}$, for all $n \geq 4$. At the same time we will show that there exists a (B)-enumeration for all $n \geq 3$.

Lemma 1 *Assume that there exist a* (B)*-enumeration* $\sigma_1, \dots, \sigma_{(n-1)!}$ *of permutations of the set* $\{1, \dots, n-1\}$. *Then there exists an* (A)*-enumeration* $\tau_1, \dots, \tau_{n!}$ *of permutations of the set* $\{1, \dots, n\}$.

Proof Let $\tau_{n(l-1)+1} = (\sigma_l(1), \dots, \sigma_l(n-1), n)$ for all $1 \leq l \leq (n-1)!$, and permutations with indices from $n(l-1)+2$ to nl are obtained by subsequent left circular shifts by one position of the permutation $\tau_{n(l-1)+1}$. Then $\tau_1, \dots, \tau_{n!}$ is an (A)-enumeration. Indeed, the condition holds by the construction for all adjacent permutations except, possibly, permutations with indices nl and $nl + 1$, $1 \leq l < (n-1)!$, or indices $n!$ and 1. But the permutation τ_{nl} has a form $(n, \sigma_l(1), \sigma_l(2), \dots, \sigma_l(n-1))$, and the permutation τ_{nl+1} has a form $(\sigma_{l+1}(1), \sigma_{l+1}(2), \ldots, \sigma_{l+1}(n-1), n)$. Therefore, the condition holds for those permutations as well, because the permutations σ_l and σ_{l+1} are adjacent under the (B)-enumeration. □

Lemma 2 *Assume that $n \geq 4$ and there exist an (A)-enumeration $\sigma_1, \ldots, \sigma_{(n-1)!}$ of permutations of the set $\{1, \ldots, n-1\}$. Then there exists a (B)-enumeration $\tau_1, \ldots, \tau_{n!}$ of permutations of the set $\{1, \ldots, n\}$.*

Proof Let $\tau_{n(l-1)+1} = (n, \sigma_l(1), \ldots, \sigma_l(n-1))$ for all $1 \leq l \leq (n-1)!$, and permutations τ_{nl}, $\tau_{n(l-1)+2}$, $\tau_{n(l-1)+3}$, ..., τ_{nl-1} (exactly in this order) are obtained by subsequent left circular shifts by one position of the permutation $\tau_{n(l-1)+1}$. Then $\tau_1, \ldots, \tau_{n!}$ is a (B)-enumeration. Indeed, the condition holds by the construction for all adjacent permutations except, possibly, permutations with indices nl and $nl+1$, $1 \leq l < (n-1)!$, or indices $n!$ and 1. But the permutations τ_{nl} and τ_{nl+1} have a form $(\sigma_l(1), \sigma_l(2), \ldots, \sigma_l(n-1), n)$ and $(n, \sigma_{l+1}(1), \sigma_{l+1}(2), \ldots, \sigma_{l+1}(n-1))$, respectively. Hence the condition holds for those permutations as well, because the permutations σ_l and σ_{l+1} are adjacent under the (A)-enumeration. □

Notice that for both $n = 3$ and $n = 4$, there exist (B)-enumerations of permutations of the set $\{1, \ldots, n\}$:

$$(1,2,3), (3,2,1), (2,1,3), (1,3,2), (2,3,1), (3,1,2);$$

$$\begin{gathered}(1,2,3,4), (2,3,4,1), (3,4,1,2), (4,1,2,3), (3,1,2,4), (1,2,4,3),\\ (2,4,3,1), (4,3,1,2), (2,3,1,4), (3,1,4,2), (1,4,2,3), (4,2,3,1),\\ (2,1,3,4), (1,3,4,2), (3,4,2,1), (4,2,1,3), (3,2,1,4), (2,1,4,3),\\ (1,4,3,2), (4,3,2,1), (1,3,2,4), (3,2,4,1), (2,4,1,3), (4,1,3,2).\end{gathered}$$

Therefore, by Lemma 1 for $n = 4$ there exists an (A)-enumeration of permutations of the set $\{1, \ldots, n\}$, and further due to Lemmas 1 and 2 we get by induction on n that there exist (A)- and (B)-enumerations of permutations of the set $\{1, \ldots, n\}$ for all $n \geq 4$.
Answer: $n = 2$ or $n \geq 4$.

15 The function $f(a) = \int_{\mathbb{R}} xe^{ax}\,d\mu(x)$ is well-defined because

$$|xe^{ax}| \leq (e^x + e^{-x})e^{ax} = e^{(a+1)x} + e^{(a-1)x}.$$

By theorem on the continuity of a Lebesgue integral with respect to a parameter, it holds that $f \in C(\mathbb{R})$. For some $\varepsilon > 0$, we have $\mu([\varepsilon; +\infty)) > 0$. On the right-hand side of the equality

$$f(a) = \int_{(-\infty;0)} xe^{ax}\,d\mu(x) + \int_{[0,+\infty)} xe^{ax}\,d\mu(x),$$

the first summand tends to 0, as $a \to +\infty$, by Lebesgue's dominated convergence theorem, and the second summand is greater than or equal to $\varepsilon e^{a\varepsilon}\mu([\varepsilon; +\infty))$, which tends to infinity as $a \to +\infty$. Hence $f(a) \to +\infty$, as $a \to +\infty$. In a similar way $f(a) \to -\infty$, as $a \to -\infty$. By the intermediate value theorem, there exists a_0 such that $f(a_0) = 0$. For $a_1 < a_2$ and all $x \in \mathbb{R} \setminus \{0\}$, it holds $x\,(e^{a_2 x} - e^{a_1 x}) > 0$, whence

$$f(a_2) - f(a_1) = \int_{\mathbb{R}} x\left(e^{a_2 x} - e^{a_1 x}\right) d\mu(x) > 0.$$

Thus, f is increasing and a required value a_0 is unique.

16 Denote by τ_i the number of members equal to i in the sequence $\{x_n,\ n \geq 0\}$. Then $\tau_i,\ n \geq 0$ are independent identically distributed random variables. It can be verified by straightforward calculation of probabilities $\mathsf{P}(\tau_0 = k_0, \ldots, \tau_n = k_n)$. Furthermore,

$$\mathsf{P}(\tau_0 = k) = p^{k-1}(1-p),\ k \geq 1.$$

Define $S_l = \sum_{j=0}^{l} \tau_j$, $l \geq 0$, and $\sigma_n = \max\{k : S_k \leq n\}$, $n \geq 1$. From the strong law of large numbers it follows that $\frac{S_l}{l} \to \mathsf{E}\tau_0$, almost surely, as $l \to \infty$. It implies that $\frac{\sigma_n}{n} \xrightarrow{\mathsf{P1}} (\mathsf{E}\tau_0)^{-1}$, as $n \to \infty$. Now,

$$\frac{1}{n}\sum_{k=0}^{n} \xi_{x_k} = \frac{1}{n}\sum_{l=0}^{\sigma_n} \xi_l \tau_l + \frac{1}{n} r_n,$$

where $|r_n| \leq |\xi_{\sigma_n+1}|\,\tau_{\sigma_n+1}$. Since $\mathsf{E}|\xi_0| < +\infty$, we get $\frac{1}{n} r_n \xrightarrow{\mathsf{P1}} 0$, as $n \to \infty$. Due to the independence of ξ and τ and the strong law of large numbers, we have

$$\frac{1}{n}\sum_{l=0}^{\sigma_n} \xi_l \tau_l = \frac{\sigma_n}{n} \cdot \frac{1}{\sigma_n}\sum_{l=0}^{\sigma_n} \xi_l \tau_l \xrightarrow{\mathsf{P1}} 0, \text{ as } n \to \infty.$$

17 Denote $S_k = \sum_{i=1}^{k} X_i$, and $T_k = \sum_{i=1}^{k} X_i^2$, $1 \leq k \leq 2n$. Then it holds

$$\begin{aligned}
\mathsf{E}Y_{2n}^2 &= \mathsf{E}\frac{S_{2n}^2}{T_{2n}} = 1 + 2n(2n-1)\mathsf{E}\frac{X_1 X_n}{T_{2n}} = 1 + \frac{2n(2n-1)}{n^2}\mathsf{E}\frac{S_n(S_{2n}-S_n)}{T_{2n}} \leq \\
&\leq 1 + 4\mathsf{E}\frac{|S_n||S_{2n}-S_n|}{T_{2n}} \leq 1 + 4\mathsf{E}\frac{|S_n||S_{2n}-S_n|}{\sqrt{T_n}\sqrt{T_{2n}-T_n}} = \\
&= 1 + 4\mathsf{E}\frac{|S_n|}{\sqrt{T_n}} \cdot \mathsf{E}\frac{|S_{2n}-S_n|}{\sqrt{T_{2n}-T_n}} = 1 + 4\left(\mathsf{E}\frac{|S_n|}{\sqrt{T_n}}\right)^2 = 1 + 4\,(\mathsf{E}Y_n)^2.
\end{aligned}$$

2010

1 Consider piecewise linear functions $F_\alpha : [0, 1] \to \mathbb{R}$, $\alpha \in (0, 1)$, defined as

$$F_\alpha(x) = \begin{cases} \frac{x}{\alpha}, & 0 \le x \le \alpha; \\ \frac{1-x}{1-\alpha}, & \alpha < x \le 1. \end{cases}$$

The graphs of all the functions F_α pass through the points $(0, 0)$ and $(1, 0)$. For $\alpha < \beta$, the graphs of the functions F_α and F_β, besides these two points, intersect at a single point $\left(\frac{\beta}{1-\alpha+\beta}, \frac{1}{1-\alpha+\beta}\right)$. Hence for every $\alpha < \beta < \gamma$, the graphs of the functions F_α, F_β, and F_γ have exactly two common points. It remains to put $f_\alpha = F_{g(\alpha)}$, $\alpha \in \mathbb{R}$, where $g : \mathbb{R} \to (0, 1)$ is an arbitrary bijection.
Answer: yes, there exists.

2 Introduce a sequence $b_n = \frac{a_n}{n}$, $n \ge 1$. By the condition of the problem

$$[n(n+1)b_n] = [n(n+1)b_{n+1}], \ n \ge 1.$$

Therefore, $|n(n+1)b_n - n(n+1)b_{n+1}| < 1$, i.e., $|b_n - b_{n+1}| < \frac{1}{n(n+1)}$, $n \ge 1$. Hence for all $n \ge 1$, $p \ge 1$, it holds

$$|b_{n+p}-b_n| \le |b_{n+1} - b_n| + \ldots + |b_{n+p} - b_{n+p-1}| <$$
$$< \frac{1}{n(n+1)} + \ldots + \frac{1}{(n+p-1)(n+p)} = \frac{1}{n} - \frac{1}{n+p} < \frac{1}{n} \to 0, \text{ as } n \to \infty.$$

Thus, $\{b_n\}$ is a Cauchy sequence, so it converges. Denote $c = \lim_{n\to\infty} b_n$. For every $n \ge 1$, we have $|b_n - c| = \lim_{p\to\infty} |b_{n+p} - b_n| \le \frac{1}{n}$. Hence

V. Brayman and A. Kukush, *Undergraduate Mathematics Competitions (1995–2016)*, Problem Books in Mathematics,
DOI 10.1007/978-3-319-58673-1_38

$$|b_n - c| \le |b_n - b_{n+1}| + |b_{n+1} - c| < \frac{1}{n(n+1)} + \frac{1}{n+1} = \frac{1}{n}.$$

Therefore, $\left|\frac{a_n}{n} - c\right| < \frac{1}{n}$, i.e., $|a_n - cn| < 1,\ n \ge 1$.

Alternative solution. For each $n \ge 1$, the inequality $|a_n - cn| < 1$ is equivalent to the double inequality

$$c_n := \frac{a_n - 1}{n} < c < C_n := \frac{a_n + 1}{n}.$$

Show that all the intervals (c_n, C_n), $n \ge 1$, have a common point. The condition of the problem implies that $|na_{n+1} - (n+1)a_n| < 1$, $n \ge 1$, hence

$$c_{n+1} - c_n = \frac{a_{n+1} - 1}{n+1} - \frac{a_n - 1}{n} = \frac{na_{n+1} - (n+1)a_n + 1}{n(n+1)} > 0,$$

$$C_{n+1} - C_n = \frac{a_{n+1} + 1}{n+1} - \frac{a_n + 1}{n} = \frac{na_{n+1} - (n+1)a_n - 1}{n(n+1)} < 0.$$

Thus, $[c_{n+1}, C_{n+1}] \subset [c_n, C_n]$, $n \ge 1$, moreover

$$C_n - c_n = \frac{2}{n} \to 0, \text{ as } n \to \infty.$$

Therefore, by the nested intervals theorem there exists a unique point c that belongs to all the segments $[c_n, C_n]$. Finally, since

$$c \in [c_{n+1}, C_{n+1}] \subset (c_n, C_n),$$

the number c belongs to all the intervals (c_n, C_n), $n \ge 1$, as well.

3 Introduce a function

$$h(x) = f(x)(ag(b) - bg(a)) - g(x)(af(b) - bf(a)) - \left(\frac{f}{g}\right)(x)(a - b)g(a)g(b),\ x \in [a, b].$$

The function h is continuous on $[a, b]$ and differentiable on (a, b), and moreover $h(a) = h(b) = bf(a)g(b) - af(b)g(a)$. Therefore, by Rolle's theorem there exists $x \in (a, b)$ such that

$$h'(x) = f'(x)(ag(b) - bg(a)) - g'(x)(af(b) - bf(a)) - \left(\frac{f}{g}\right)'(x)(a - b)g(a)g(b) = 0.$$

4 For $k = 1$, all the elements of the set $S(A_1)$ are polynomials of A_1, therefore, each two of them commute. Hence there is no matrix A_1 for which $S(A_1) = M_n(\mathbb{R})$.

Give an example of required matrices for $k = 2$. Denote by I_{ij} the $n \times n$ matrix, the only nonzero entry of which equals 1 and is located in ith row and jth column. Put

$$A_1 = I_{11}, \quad A_2 = \begin{pmatrix} 0 & 1 & 0 & \dots & 0 \\ 0 & 0 & \dots & \dots & 0 \\ \dots & \dots & \dots & \dots & \dots \\ 0 & 0 & 0 & \dots & 1 \\ 1 & 0 & 0 & \dots & 0 \end{pmatrix}$$

(here A_2 is a permutation matrix).

It is easy to verify that $A_2^{n-i+1} A_1 A_2^{n-j+1} = I_{ij}$. Thus, the set $S(A_1, A_2)$ contains all the matrices I_{ij}, $1 \le i, j \le n$, which form a basis in $M_n(\mathbb{R})$. Hence $S(A_1, A_2) = M_n(\mathbb{R})$.

Answer: $k = 2$.

5 The sequence $\{x_n\}$ is increasing, hence it has a finite or infinite limit. If $\lim_{n\to\infty} x_n = x < +\infty$, then $x = x + e^{-x}$, a contradiction. Therefore, $\lim_{n\to\infty} x_n = +\infty$. We use twice the Stolz–Cesaro theorem and obtain

$$\lim_{n\to\infty} \frac{x_n}{\ln n} = \lim_{n\to\infty} \frac{x_{n+1} - x_n}{\ln(n+1) - \ln n} = \lim_{n\to\infty} \frac{e^{-x_n}}{\ln(1+\frac{1}{n})} = \lim_{n\to\infty} \frac{n}{e^{x_n}} = \lim_{n\to\infty} \frac{1}{e^{x_{n+1}} - e^{x_n}}.$$

It remains to notice that

$$\lim_{n\to\infty} (e^{x_{n+1}} - e^{x_n}) = \lim_{n\to\infty} (e^{x_n + e^{-x_n}} - e^{x_n}) = \lim_{n\to\infty} \frac{e^{e^{-x_n}} - 1}{e^{-x_n}} = \lim_{t\to 0} \frac{e^t - 1}{t} = 1.$$

Answer: 1.

6 We check that for each polynomial $P(x, y)$, there exists a polynomial $Q(x, y)$ of a form

$$\sum_{i,j=0}^{n-1} c_{ij} x^i y^j \qquad (*)$$

such that $\deg Q \le \deg P$ and $Q(i, j) = P(i, j)$ for all $1 \le i, j \le n$. Indeed, one can rewrite $P(x, y)$ as $\sum_{l\ge 0} P_l(x) y^l$ and replace each polynomial $P_l(x)$ with its remainder after division by $\prod_{i=1}^{n} (x - i)$, and then rewrite the result as $\sum_{l=0}^{n-1} x^l \widetilde{P}_l(y)$ and replace each polynomial $\widetilde{P}_l(y)$ with its remainder after division by $\prod_{j=1}^{n} (y - j)$. Thus, it suffices to consider only the polynomials $P(x, y)$ of the form $(*)$.

Notice that for arbitrary numbers a_{ij}, there exists an interpolating polynomial of the form $(*)$, namely

$$P(x,y)=\sum_{i_0,j_0=1}^{n} a_{i_0j_0}\cdot\prod_{\substack{i=0\\ i\neq i_0}}^{n}\frac{x-i}{i_0-i}\cdot\prod_{\substack{j=0\\ j\neq j_0}}^{n}\frac{y-j}{j_0-j}. \qquad (**)$$

Thus, $k \le 2n-2$. On the other hand, both the set of polynomials of the form $(*)$ and the set of collections of real numbers $\{a_{ij},\ 1\le i,j\le n\}$ are real linear spaces of the same dimension n^2. Hence, the relation $(**)$ determines a one-to-one correspondence between the collections of numbers $\{a_{ij}\}$ and the polynomials of the form $(*)$. For instance, a unique polynomial of the form $(*)$, which corresponds to the numbers $a_{ij}\equiv i^{n-1}j^{n-1}$, is equal to $x^{n-1}y^{n-1}$ and has a degree $2n-2$. Therefore, $k=2n-2$.
Answer: $k=2n-2$.

7 Denote by A_{ij} the matrix obtained from A by erasing of ith row and jth column, $1\le i,j\le n$. In particular $A_{ii}=A_i$. Since $\det A=0$ and $\det A_1=2010\ne 0$, the last $n-1$ columns of the matrix A are linearly independent and the first column is their linear combination with some coefficients $\lambda_2,\dots,\lambda_n$. Because A is symmetric, the first row is a linear combination of the last $n-1$ rows with the same coefficients $\lambda_2,\dots,\lambda_n$, and since A has integer entries, it holds $\lambda_2,\dots,\lambda_n\in\mathbb{Q}$. Fix $i>1$ and consider the matrix A_{i1}. Its first row is a linear combination of other rows of this matrix and of ith row of A with deleted first entry. Hence it is easy to verify that $\det A_{i1}=(-1)^{i-1}\lambda_i\det A_{11}$. In a similar way

$$\det A_i=\det A_{ii}=(-1)^{i-1}\lambda_i\det A_{i1}=\lambda_i^2\det A_{11}=2010\lambda_i^2.$$

Notice that $\det A_i=2010\lambda_i^2$ is an integer as the determinant of a matrix with integer entries. If λ_i is not an integer, then λ_i equals to an irreducible fraction $\frac{s}{t}$ with $s\in\mathbb{Z}$ and $t\in\mathbb{N}$, $t>1$. Then $2010=2\cdot3\cdot5\cdot67$ is not divisible by t^2, hence $2010\lambda_i^2=\frac{2010s^2}{t^2}$ is not an integer, a contradiction. Therefore, λ_i is an integer and $\det A_i=2010\lambda_i^2$ is divisible by 2010.

Alternative solution. The condition of the problem implies that $\operatorname{rk} A=n-1$. Let $\widetilde{A}=(\widetilde{a}_{ij})_{i,j=1}^n$ be the adjugate matrix, that is $\widetilde{a}_{ij}=(-1)^{i+j}\det A_{ji}$, where A_{ij} is the matrix obtained from A by erasing of ith row and jth column, $1\le i,j\le n$. Then $A\widetilde{A}=\det A\cdot I=0$, where I is the identity $n\times n$ matrix. Well-known inequality $\operatorname{rk} AB\ge \operatorname{rk} A+\operatorname{rk} B-n$ implies that

$$0=\operatorname{rk} A\widetilde{A}\ge \operatorname{rk} A+\operatorname{rk}\widetilde{A}-n=\operatorname{rk}\widetilde{A}-1.$$

Hence $\operatorname{rk}\widetilde{A}\le 1$, i.e., all the rows of $\widetilde{A}$ are proportional to each other. Let $\widetilde{a}_{i1}=\mu_i\widetilde{a}_{11}$. Then $\widetilde{a}_{ii}=\mu_i\widetilde{a}_{i1}=\mu_i^2\widetilde{a}_{11}$. But all $\widetilde{a}_{ij}$ are integers, hence $\mu_i\in\mathbb{Q}$. Now,

$$\det A_i=\widetilde{a}_{ii}=\mu_i^2\widetilde{a}_{11}=2010\mu_i^2,$$

and the solution can be finished similar to the previous one.

8 Suppose that the linear span of vectors $\overrightarrow{v}_{ij}$, $1 \le i < j \le n$, does not coincide with $\mathbb{R}^n$. Then there exists a nonzero vector $\overrightarrow{x} = (x_1, \ldots, x_n)$ orthogonal to all the vectors $\overrightarrow{v}_{ij}$, that is

$$\forall\, 1 \le i < j \le n \qquad (\overrightarrow{x}, \overrightarrow{v}_{ij}) = \sum_{k=1}^{n} x_k \delta_{ijk} = 0.$$

Therefore, one can place the numbers $x_1, \ldots, x_n$ at the points $P_1, \ldots, P_n$ in such a way that the sum of numbers on each line P_iP_j equals zero. Denote $S = x_1 + \ldots + x_n$. Fix an arbitrary point P_i, $1 \le i \le n$, and consider all the lines P_iP_j, $j \ne i$. Assume that there are exactly k_i distinct lines among them. Notice that $k_i > 1$ by the condition of the problem. The sum of numbers on each lines P_iP_j equals zero. The sum of these sums over k_i distinct lines equals $S + (k_i - 1)x_i = 0$, because the summand x_i is counted k_i times, and each of the summands x_j, $j \ne i$, is counted once. Hence $x_i = -\frac{S}{k_i - 1}$, $1 \le i \le n$. Thus, the coordinates of nonzero vector $\overrightarrow{x}$ are either all positive or all negative. Therefore, the sum of numbers on each line P_iP_j is nonzero, a contradiction.

Alternative solution. Assume that there are exactly m distinct lines among the lines P_iP_j. Introduce $n \times m$ matrix A, columns of which are vectors $\overrightarrow{v}_{ij}$ corresponding to distinct lines. Set $B = AA^{\mathrm{T}} = (b_{ij})_{i,j=1}^n$. It is not difficult to verify that b_{ij} is the number of lines passing through both points P_i and P_j. Thus, $b_{ij} = 1$ for $i \ne j$, and $b_{ii} = k_i$, where $k_i > 1$ is the number of lines passing through the point P_i, $1 \le i \le n$. If the determinant of B is nonzero, then $\operatorname{rk} A \ge \operatorname{rk} B = n$, and the linear span of the columns of A coincides with $\mathbb{R}^n$. It remains to compute $\det B$. Subtract the first row from all the rest ones, and then subtract from the first row all the other rows multiplied by $\frac{1}{k_2 - 1}, \ldots, \frac{1}{k_n - 1}$, respectively. We get

$$\det B = \begin{vmatrix} k_1 & 1 & 1 & \ldots & 1 \\ 1 & k_2 & 1 & \ldots & 1 \\ 1 & 1 & k_3 & \ldots & 1 \\ \ldots & \ldots & \ldots & \ldots & \ldots \\ 1 & 1 & 1 & \ldots & k_n \end{vmatrix} = \begin{vmatrix} k_1 & 1 & 1 & \ldots & 1 \\ 1-k_1 & k_2-1 & 0 & \ldots & 0 \\ 1-k_1 & 0 & k_3-1 & \ldots & 0 \\ \ldots & \ldots & \ldots & \ldots & \ldots \\ 1-k_1 & 0 & 0 & \ldots & k_n-1 \end{vmatrix} =$$

$$= \begin{vmatrix} k_1 + (k_1 - 1)\left(\frac{1}{k_2-1} + \ldots + \frac{1}{k_n-1}\right) & 0 & 0 & \ldots & 0 \\ 1-k_1 & k_2-1 & 0 & \ldots & 0 \\ 1-k_1 & 0 & k_3-1 & \ldots & 0 \\ \ldots & \ldots & \ldots & \ldots & \ldots \\ 1-k_1 & 0 & 0 & \ldots & k_n-1 \end{vmatrix} =$$

$$= \left(k_1 + (k_1 - 1)\left(\frac{1}{k_2-1} + \ldots + \frac{1}{k_n-1}\right)\right)(k_2 - 1) \cdot \ldots \cdot (k_n - 1) =$$

$$= (k_1 - 1)(k_2 - 1) \cdot \ldots \cdot (k_n - 1)\left(1 + \frac{1}{k_1-1} + \frac{1}{k_2-1} + \ldots + \frac{1}{k_n-1}\right) > 0,$$

which finishes the proof.

9 Introduce an ellipse E_1 with foci $(x, 2-x)$ and $(2-x, x)$, $0 < x < 1$, that touches coordinate axes at the points $(x^2 - 2x + 2, 0)$ and $(0, x^2 - 2x + 2)$, and an ellipse E equal to E_1 with foci $\left(-\sqrt{2}(1-x), 0\right)$ and $\left(\sqrt{2}(1-x), 0\right)$. If $x = 0$, the ellipses becomes a couple of disjoint segments, and if $x = 1$, the ellipses becomes a couple of intersecting circles. It is not difficult to verify that the minimal distance between the points of E_1 and E is a continuous function of x. Denote by x_* the infimum of $0 < x < 1$ for which the ellipses intersect. By the continuity, for $x = x_*$ the ellipses E_1 and E touch. Reflect the ellipse E_1 with respect to the axes and the origin to obtain three more required ellipses E_2, E_3, and E_4 (see Fig. 1).

Suppose that there exist equal ellipses E, E_1, E_2, E_3 that pairwise touch each other. It is easy to verify that one of the ellipses lies inside a triangle with vertices at the tangent points of three other ellipses. But then the major axis of this ellipse is less than the largest side of the triangle, which is a chord of another ellipse. Thus, the major axes of ellipses cannot be equal, a contradiction.

Answer: $N = 4$.

Fig. 1 Touching ellipses

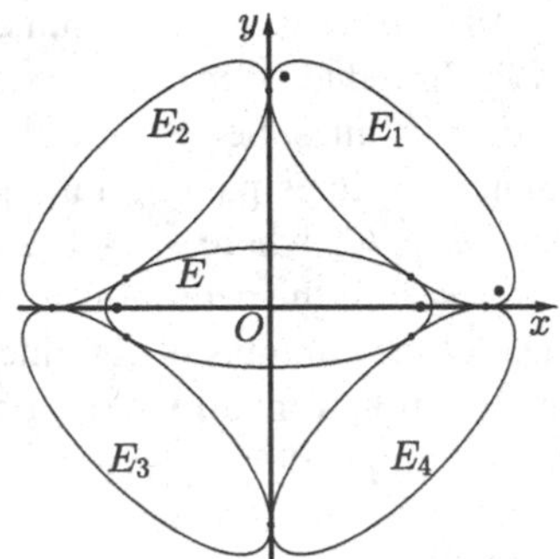

10 For every $0 < \varepsilon < 1$, there exists $C = C(\varepsilon)$ such that for all $x \geq C$, it holds

$$\frac{1-\varepsilon}{x^3} < f(x) < \frac{1+\varepsilon}{x^3}.$$

Hence for all $c \geq C$, the integral $\int\limits_c^{+\infty} (x-c) f(x)\, dx$ is well defined, moreover

$$\frac{1-\varepsilon}{2} = (1-\varepsilon)c \int\limits_c^{+\infty} \frac{x-c}{x^3}\, dx \leq c \int\limits_c^{+\infty} (x-c) f(x)\, dx \leq (1+\varepsilon)c \int\limits_c^{+\infty} \frac{x-c}{x^3}\, dx = \frac{1+\varepsilon}{2}.$$

Since $0 < \varepsilon < 1$ is arbitrary, the limit in question equals $\frac{1}{2}$.

Answer: $\frac{1}{2}$.

11 By Jensen's inequality $f(\alpha \sin x) \le \frac{1+\alpha}{2} f(\sin x) + \frac{1-\alpha}{2} f(-\sin x)$. Therefore,

$$\int_{-\pi/2}^{\pi/2} f(\alpha \sin x)dx \le$$

$$\le \frac{1+\alpha}{2} \int_{-\pi/2}^{\pi/2} f(\sin x)dx + \frac{1-\alpha}{2} \int_{-\pi/2}^{\pi/2} f(-\sin x)dx = \int_{-\pi/2}^{\pi/2} f(\sin x)dx.$$

12 Introduce a set $A = \{(x, f(x)) \mid x \in \mathbb{R}\} \subset \mathbb{R}^2$. It is Borel measurable because

$$A = \{(x, y) \mid y - f(x) \in \{0\}\} = \varphi^{-1}(\{0\}),$$

where the set $\{0\}$ is closed, and the function $\varphi(x, y) = y - f(x) : \mathbb{R}^2 \to \mathbb{R}$ is Borel measurable as a difference of continuous function $\pi_2(x, y) = y$ and the function $f(\pi_1(x, y)) = f(x)$, which is a composition of continuous function $\pi_1(x, y) = x$ and Borel measurable function f. We have

$$\mathsf{P}(\{(\xi, \eta) \in A\}) = \mathsf{P}(\{(\xi, f(\xi)) \in A\}) = 1,$$

i.e., $(\xi, \eta) \in A$ almost surely. Hence $\eta = f(\xi)$ almost surely.

15 Define

$$\|x\| = \sqrt{\left(x_1 - \sum_{k=2}^{\infty} k^2 x_k\right)^2 + x_2^2 + x_3^2 + \ldots + x_k^2 + \ldots}, \; x \in X.$$

It is not difficult to verify that it is indeed a norm on X. Consider the sequence

$$x(n) = \left(n, \underbrace{0, \ldots, 0}_{n-2}, \frac{1}{n}, 0, \ldots\right), n \ge 2.$$

Then $\|x(n)\| = \frac{1}{n} \to 0$, as $n \to \infty$, but $f(x(n)) = n \to \infty$, as $n \to \infty$. Thus, f is discontinuous.
Answer: yes, there exists.

Remark 1 One can give other examples of a required norm:

$$\|x\| = \sqrt{\int_0^1 \left(\sum_{k=1}^{\infty} x_k t^{k-1}\right)^2 dt}, \; x \in X; \quad \|x\| = \int_0^1 \left|\sum_{k=1}^{\infty} x_k t^{k-1}\right| dt, \; x \in X.$$

18 The Gram determinant $\Gamma(\xi_1, \ldots, \xi_n)$ is equal to a squared volume of a parallelepiped determined by the vectors $\xi_1, \ldots, \xi_n$. Hence

$$\Gamma(\xi_1, \ldots, \xi_n) = \|\xi_1\|^2 \cdot d_2^2 \cdot \ldots \cdot d_n^2,$$

where d_k is the length of component of ξ_k which is orthogonal to the subspace generated by the vectors $\xi_1, \ldots, \xi_{k-1}$.

Prove that for each $1 \le k \le n$,

$$\mathsf{P}\left(\Gamma(\xi_1, \ldots, \xi_k) \ne 0\right) = 1 \text{ and } \mathsf{E}\Gamma(\xi_1, \ldots, \xi_k) = n(n-1) \cdot \ldots \cdot (n-k+1).$$

For $k = 1$, these statements are obvious. Assume that they hold for some k, $1 \le k \le n-1$. Consider $\Gamma(\xi_1, \ldots, \xi_{k+1}) = \Gamma(\xi_1, \ldots, \xi_k) \cdot d_{k+1}^2$. By the inductive hypothesis $\xi_1, \ldots, \xi_k$ are linearly independent with probability 1. Therefore, for fixed $\xi_1, \ldots, \xi_k$, the random variable d_{k+1}^2 has χ_{n-k}^2 distribution, i.e., the same distribution as a squared norm of a Gaussian vector of dimension $n-k$ with zero mean and unit covariance matrix. Thus, $d_{k+1}^2 \ne 0$ almost surely, and $\mathsf{P}\left(\Gamma(\xi_1, \ldots, \xi_{k+1}) \ne 0\right) = 1$. Next,

$$\begin{aligned}\mathsf{E}\Gamma(\xi_1, \ldots, \xi_{k+1}) =& \mathsf{E}(\mathsf{E}\Gamma(\xi_1, \ldots, \xi_{k+1})/\xi_1, \ldots, \xi_k) = \\ =& \mathsf{E}\Gamma(\xi_1, \ldots, \xi_k)\mathsf{E}(d_{k+1}^2/\xi_1, \ldots, \xi_k) = \\ =& \mathsf{E}\Gamma(\xi_1, \ldots, \xi_k)(n-k) = n(n-1) \cdot \ldots \cdot (n-k).\end{aligned}$$

By the principle of mathematical induction the statement is proved. For $k = n$, it follows that $\mathsf{E}\Gamma(\xi_1, \ldots, \xi_n) = n!$

Alternative solution. Let $\xi_i = (\xi_{i1}, \xi_{i2}, \ldots, \xi_{in})$, $1 \le i \le n$. Put $A = (\xi_{ij})_{i,j=1}^n$. Then $AA^{\mathrm{T}} = \left((\xi_i, \xi_j)\right)_{i,j=1}^n$, therefore, $\Gamma(\xi_1, \ldots, \xi_n) = \det(AA^{\mathrm{T}}) = (\det A)^2$. Hence

$$\begin{aligned}\mathsf{E}\Gamma(\xi_1, \ldots, \xi_n) =& \mathsf{E}(\det A)^2 = \mathsf{E}\left(\sum_{\sigma \in S_n} (-1)^{\operatorname{sgn}(\sigma)} \xi_{1\sigma(1)} \cdot \ldots \cdot \xi_{n\sigma(n)}\right)^2 = \\ =& \sum_{\sigma,\tau \in S_n} (-1)^{\operatorname{sgn}(\sigma)+\operatorname{sgn}(\tau)} \mathsf{E}(\xi_{1\sigma(1)} \cdot \ldots \cdot \xi_{n\sigma(n)} \xi_{1\tau(1)} \cdot \ldots \cdot \xi_{n\tau(n)}).\end{aligned}$$

Since zero mean random vectors $\xi_1, \ldots, \xi_n$ are stochastically independent and have unit covariance matrix, it holds

$$\begin{aligned}\mathsf{E}(\xi_{1\sigma(1)} \cdot \ldots \cdot \xi_{n\sigma(n)} \xi_{1\tau(1)} \cdot \ldots \cdot \xi_{n\tau(n)}) =& \\ =& \mathsf{E}(\xi_{1\sigma(1)}\xi_{1\tau(1)}) \cdot \ldots \cdot \mathsf{E}(\xi_{n\sigma(n)}\xi_{n\tau(n)}) = \begin{cases} 1, & \sigma = \tau, \\ 0, & \sigma \ne \tau, \end{cases}\end{aligned}$$

therefore, $\mathsf{E}\Gamma(\xi_1, \ldots, \xi_n) = \sum_{\sigma = \tau \in S_n} 1 = n!$

Answer: $n!$

2011

1 Let $f_1(x) = x^2$ and $f_2(x) = \frac{1}{2}(x^2 + g(x))$, $x \in [0, 1]$, where the function g is defined as follows: $g(0) = 0$, $g\left(\frac{1}{n}\right) = \frac{1}{n^2}$, $n \geq 1$, and g is linear on the segment $\left[\frac{1}{n+1}, \frac{1}{n}\right]$ for each $n \geq 1$. It is not difficult to verify that g is convex, hence the functions f_1 and f_2 are strictly convex and satisfy the condition of the problem.
Answer: yes, there exist.

2 Since there exists a continuous inverse map $\varphi^{-1} : \mathbb{R} \to \mathbb{R}$, the equality $\varphi^{(n)}(t) = f(\varphi(t))$, $t \in \mathbb{R}$, holds true for a unique continuous function $f(t) = \varphi^{(n)}(\varphi^{-1}(t))$.

3 Notice that a symmetric matrix always has an eigenbasis. Let $\lambda_1, \ldots, \lambda_n > 0$ be eigenvalues of the matrix A. Then $\lambda_1\lambda_2 \ldots \lambda_n = \det A \geq 1$, because the determinant of A is a positive integer. On the other hand, $\lambda_1 + \lambda_2 + \ldots + \lambda_n = \operatorname{tr} A \leq n$. By the AM-GM inequality it holds

$$1 \geq \frac{\lambda_1 + \ldots + \lambda_n}{n} \geq \sqrt[n]{\lambda_1 \ldots \lambda_n} \geq 1,$$

hence the equality is attained in the AM-GM inequality. Therefore, $\lambda_1 = \ldots = \lambda_n = 1$ and A is the identity matrix.

Alternative solution. Show that the matrix A is diagonal. Indeed, suppose that $a_{ij} \neq 0$ for some $i \neq j$. Then $a_{ij} = a_{ji} = 1$. Introduce a vector $\overrightarrow{v} = \overrightarrow{e_i} - \overrightarrow{e_j}$, where $\overrightarrow{e_1}, \ldots, \overrightarrow{e_n}$ is a standard basis in $\mathbb{R}^n$. Then $\overrightarrow{v}^{\mathrm{T}} A \overrightarrow{v} = a_{ii} + a_{jj} - 2 \leq 0$, a contradiction because the matrix A is positive definite. Since the eigenvalues of A are positive, there are no zeros on its main diagonal, and finally A is the identity matrix.
Answer: A is the identity matrix.

4 Let the ellipses $e(A, B, C)$, $e(B, C, A)$, and $e(C, A, B)$ share a common point D. Then

V. Brayman and A. Kukush, *Undergraduate Mathematics Competitions (1995–2016)*, Problem Books in Mathematics,
DOI 10.1007/978-3-319-58673-1_39

$$\begin{cases} AD + BD = AC + BC, \\ BD + CD = AB + AC, \\ AD + CD = AB + BC, \end{cases}$$

whence $AD = BC$, $BD = AC$, $CD = AB$. Therefore, the triangles ABC, DCB, CDA, and BAD are equal. Let $A_0B_0C_0$ be the triangle for which ABC is the medial triangle. Due to the obtained equalities D is either a vertex of the triangle $A_0B_0C_0$, or it belongs simultaneously to all its sides, which is impossible. Without loss of generality assume that D and B_0 coincide. Then $ABCD$ is a parallelogram with equal diagonals, i.e., a rectangle, whence $\angle ABC = 90°$. On the other hand, if ABC is a right triangle ($\angle ABC = 90°$), then it is evident that each of the ellipses $e(A, B, C)$, $e(B, C, A)$, and $e(C, A, B)$ passes through a vertex D of a rectangle $ABCD$, since D is symmetric to C with respect to the bisector of AB, symmetric to A with respect to the bisector of BC, and symmetric to B with respect to midpoint of AC.
Answer: ABC is an arbitrary right triangle.

5 It is easy to verify that $x = 0$, $x = 1$, and $x = 2$ satisfy the equation. Show that there are no other roots. Introduce a function

$$f(x) = 9^x + 4^x + 2^x - 8^x - 6^x - 1$$

and suppose that it has at least 4 zeros. By Rolle's theorem if a function $g(x)$ has at least n zeros $x_1 < x_2 < \ldots < x_n$, then the function $D_a g(x) := a^x(g(x)a^{-x})'$ has at least $n-1$ zeros $y_1, \ldots, y_{n-1}$, where $x_1 < y_1 < x_2 < \ldots < x_{n-1} < y_{n-1} < x_n$. Hence the function $D_8 D_6 D_1 f(x)$ should have at least one zero. But the function

$$D_8 D_6 D_1 f(x) = \ln\frac{9}{8} \cdot \ln\frac{9}{6} \cdot \ln 9 \cdot 9^x + \ln\frac{4}{8} \cdot \ln\frac{4}{6} \cdot \ln 4 \cdot 4^x + \ln\frac{2}{8} \cdot \ln\frac{2}{6} \cdot \ln 2 \cdot 2^x$$

is always positive, a contradiction.
Answer: $0, 1, 2$.

6 By the monotone convergence theorem, it is easy to show that $x_n \to 0$, as $n \to \infty$. Next, by the Stolz–Cesaro theorem it holds

$$nx_n = \frac{n}{\frac{1}{x_n}} \sim \frac{1}{\frac{1}{x_n} - \frac{1}{x_{n-1}}} = \frac{x_{n-1}x_n}{x_{n-1} - x_n} =$$

$$= \frac{x_{n-1}(x_{n-1} - x_{n-1}^2)}{x_{n-1}^2} = 1 - x_{n-1} \to 1, \text{ as } n \to \infty.$$

Transform the expression in question and use the Stolz–Cesaro theorem once again:

$$\frac{n^2x_n - n}{\ln n} = \frac{nx_n \cdot (n - \frac{1}{x_n})}{\ln n} \sim \frac{n - \frac{1}{x_n}}{\ln n} \sim \frac{1 - \frac{1}{x_n} + \frac{1}{x_{n-1}}}{\ln n - \ln(n-1)} = \frac{1 - \frac{1}{x_n} + \frac{1}{x_{n-1}}}{-\ln\left(1 - \frac{1}{n}\right)} \sim$$

$$\sim n\left(1 - \frac{1}{x_{n-1} - x_{n-1}^2} + \frac{1}{x_{n-1}}\right) = -\frac{nx_{n-1}}{1 - x_{n-1}} \to -1, \text{ as } n \to \infty.$$

Answer: -1.

7 Introduce a function $f(x) = \begin{cases} -1, & x < -1, \\ x, & -1 \le x < 2009, \\ 2009, & x \ge 2009, \end{cases}$ and check that f is 2010-positive.

Let $x_1, \ldots, x_{2010}$ be such that $x_1 + \ldots + x_{2010} \ge 0$. Notice that f is convex on $(-\infty, 2009]$. If all $x_i \le 2009$ then by Jensen's inequality it holds

$$\frac{1}{2010}\left(f(x_1) + \ldots + f(x_{2010})\right) \ge f\left(\frac{x_1 + \ldots + x_{2010}}{2010}\right) \ge f(0).$$

If there exists some $x_i \ge 2009$ then

$$\frac{1}{2010}\left(f(x_1) + \ldots + f(x_{2010})\right) \ge \frac{1}{2010}\left(2009 + 2009 \cdot (-1)\right) \ge 0 = f(0).$$

On the other side, consider $x_1 = \ldots = x_{2010} = -1$ and $x_{2011} = 2010$. We have $x_1 + \ldots + x_{2011} = 0$, but

$$\frac{1}{2011}\left(f(x_1) + \ldots + f(x_{2011})\right) = -\frac{1}{2011} < 0 = f(0).$$

Thus, f is not 2011-positive.
Answer: yes, there exists.

8 Show that

$$\lim_{n\to\infty} \frac{1}{\sqrt{n}} \sum_{k=0}^{n} \sqrt{\binom{n}{k} p^k (1-p)^{n-k}} = 0,$$

for every $0 < p < 1$ (in particular for $p = \frac{1}{2}$ we obtain the given sequence $\{a_n\}$).

Fix $0 < \delta < 1$ and consider separately index sets

$$\{0 \le k \le n : |k - np| \le \delta n\} \text{ and } \{0 \le k \le n : |k - np| > \delta n\}.$$

There exist at most $2\delta n + 1$ integers k for which $|k - np| \le \delta n$, hence by the QM-AM inequality it holds

$$\left(\sum_{|k-np|\le\delta n}\sqrt{\binom{n}{k}p^k(1-p)^{n-k}}\right)^2\le$$

$$\le(2\delta n+1)\sum_{|k-np|\le\delta n}\binom{n}{k}p^k(1-p)^{n-k}\le 2\delta n+1.$$

Again by the QM-AM inequality we get

$$\left(\sum_{|k-np|>\delta n}\sqrt{\binom{n}{k}p^k(1-p)^{n-k}}\right)^2\le n\sum_{|k-np|>\delta n}\binom{n}{k}p^k(1-p)^{n-k}=$$

$$=n\sum_{|\frac{k}{n}-p|>\delta}\binom{n}{k}p^k(1-p)^{n-k}<\frac{n}{\delta^2}\sum_{|\frac{k}{n}-p|>\delta}\binom{n}{k}p^k(1-p)^{n-k}\left(\frac{k}{n}-p\right)^2\le$$

$$\le\frac{n}{\delta^2}\sum_{k=0}^{n}\binom{n}{k}p^k(1-p)^{n-k}\left(\frac{k}{n}-p\right)^2=\frac{p(1-p)}{\delta^2}.$$

The latter equality is a corollary of well-known identities

$$\sum_{k=0}^{n}\binom{n}{k}p^k(1-p)^{n-k}=1,\quad \sum_{k=0}^{n}k\binom{n}{k}p^k(1-p)^{n-k}=np,$$

$$\sum_{k=0}^{n}k(k-1)\binom{n}{k}p^k(1-p)^{n-k}=n(n-1)p^2.$$

Thus,

$$\frac{1}{\sqrt{n}}\sum_{k=0}^{n}\sqrt{\binom{n}{k}p^k(1-p)^{n-k}}\le\sqrt{\frac{2\delta n+1}{n}}+\frac{\sqrt{p(1-p)}}{\delta\sqrt{n}}\le$$

$$\le\sqrt{2\delta}+\frac{1}{\sqrt{n}}+\frac{\sqrt{p(1-p)}}{\delta\sqrt{n}}.$$

It remains to notice that for every $\varepsilon>0$ one can choose $\delta>0$ such that $\sqrt{2\delta}<\frac{\varepsilon}{3}$, and choose $N\ge 1$ such that for all $n\ge N$ it holds $\frac{1}{\sqrt{n}}<\frac{\varepsilon}{3}$, $\frac{\sqrt{p(1-p)}}{\delta\sqrt{n}}<\frac{\varepsilon}{3}$.
Answer: 0.

9 It holds

$$\arctan(\xi+\eta)\ge\arctan\xi,\quad\text{almost surely,}\qquad(*)$$

hence $\mathsf{E} \arctan(\xi + \eta) \geq \mathsf{E} \arctan \xi$ (the expectations exist because $\arctan x$ is bounded). But $\xi + \eta$ is identically distributed with ξ, therefore, in fact the equality is attained in the latter inequality. In view of $(*)$ we have $\arctan(\xi + \eta) = \arctan \xi$ almost surely, whence $\xi + \eta = \xi$ almost surely, and $\eta = 0$ almost surely.

11 Show the following: if $g_n(x) \xrightarrow{\lambda} g(x)$, as $n \to \infty$, and a function $\varphi : \mathbb{R} \to \mathbb{R}$ is uniformly continuous on $\mathbb{R}$, then $\varphi(g_n(x)) \xrightarrow{\lambda} \varphi(g(x))$, as $n \to \infty$. Indeed, fix $\varepsilon > 0$. There exists $\delta = \delta(\varepsilon)$ such that the inequality $|t - s| \leq \delta$ implies the inequality $|\varphi(t) - \varphi(s)| < \varepsilon$. Then as $n \to \infty$, it holds

$$\lambda\left(\{x : |\varphi(g_n(x)) - \varphi(g(x))| > \varepsilon\}\right) \leq \lambda\left(\{x : |g_n(x) - g(x)| > \delta\}\right) \to 0.$$

To prove the statement of the problem consider

$$g_n(x) = (f_n(x))^{2011}, \ n \geq 1, \quad g(x) = (f(x))^{2011},$$

and notice that the function $\varphi(t) = \sqrt[2011]{t}$, $t \in \mathbb{R}$, is uniformly continuous on $\mathbb{R}$.

12 Let X be a ring with 2011 elements. Since 2011 is a prime number, the additive group of the ring is the cyclic group of order 2011. Fix a nonzero element $x \in X$. Then $X = \{0, x, 2x, \ldots, 2010x\}$. Consider two cases:
(1) $x^2 = 0$. Then for arbitrary $y, z \in X$ it holds $y = ix$, $z = jx$, whence $yz = ijx^2 = 0$, thus, it is a ring with the zero product property.
(2) $x^2 = kx$ for some integer $k \in \{1, \ldots, 2010\}$. There exists $l \in \{1, \ldots, 2010\}$ such that $kl - 1$ is divisible by 2011. Consider $a = lx$. Then $a^2 = l^2x^2 = l^2kx = lx = a$, and $X = \{0, a, 2a, \ldots, 2010a\}$, so the mapping $f : X \ni ia \mapsto i \in \mathbb{Z}_{2011}$ is an isomorphism.
Answer: there are two rings: the ring with the zero product property and $\mathbb{Z}_{2011}$.

13 See solution of the Problem 8.
Answer: 0.

14 Denote by B_i the $n \times n$ matrix the only nonzero entry of which equals 1 and is located in ith row and ith column. Put $\overrightarrow{v_1} = (\sqrt{x_1}, \sqrt{x_2}, \ldots, \sqrt{x_n})$, then $\|\overrightarrow{v_1}\| = 1$. One can extend $\overrightarrow{v_1}$ to an orthobasis $\overrightarrow{v_1}, \overrightarrow{v_2}, \ldots, \overrightarrow{v_n}$ of the space $\mathbb{R}^n$. Introduce a matrix S whose columns are vectors $\overrightarrow{v_1}, \overrightarrow{v_2}, \ldots, \overrightarrow{v_n}$. Then the matrices $A_i = S^{\mathrm{T}} B_i S$, $1 \leq i \leq n$, have required properties, because

$$A_i^2 = S^{\mathrm{T}} B_i S S^{\mathrm{T}} B_i S = S^{\mathrm{T}} B_i^2 S = S^{\mathrm{T}} B_i S = A_i,$$
$$A_i A_j = S^{\mathrm{T}} B_i S S^{\mathrm{T}} B_j S = S^{\mathrm{T}} B_i B_j S = 0, \ i \leq j,$$

and $(A_i)_{11} = x_i$.
Answer: yes, there exist.

15 The function $g(t) = t_+$ is convex on $\mathbb{R}$, hence by Jensen's inequality

$$\mathsf{E}(\xi_1 + \xi_2 + \xi_3 - K)_+ \geq (\mathsf{E}(\xi_1 + \xi_2 + \xi_3) - K)_+ = (2 - K)_+.$$

The equality is attained, e.g., under the condition $\xi_1 + \xi_2 + \xi_3 = 2$ almost surely. It remains to construct an example of random variables ξ_1, ξ_2, ξ_3 satisfying the latter condition. Let the random vector (ξ_1, ξ_2, ξ_3) be uniformly distributed in the triangle with vertices $(1, 1, 0)$, $(0, 1, 1)$, and $(1, 0, 1)$. Then $\xi_1 + \xi_2 + \xi_3 \equiv 2$ and it is not difficult to verify that each of the random variables ξ_1, ξ_2, ξ_3 has the required probability density function.
Answer: $f(K) = (2 - K)_+$, $K \geq 0$.

16 Let x_0 be a point at which the function $d(Tx, x)$ attains its minimal value on the compact set X. Suppose that x_0 is not a fixed point of the mapping T. For $x = x_0$ and $y = Tx_0$, it holds $\frac{1}{2}d(x_0, Tx_0) < d(x_0, Tx_0)$, hence $d(T^2x_0, Tx_0) < d(x, Tx_0)$, a contradiction.

Now, suppose that there exist two distinct fixed points x_0 and y_0. Then $0 = \frac{1}{2}d(x_0, Tx_0) < d(x_0, y_0)$, hence $d(x_0, y_0) < d(x_0, y_0)$, a contradiction.

2012

2012

1 Notice that $a_{ij} = -a_{ji}$, $i, j = 0, 1, \ldots, 2012$, i.e., the matrix $A = (a_{ij})_{i,j=0}^{2012}$ is skew-symmetric, $A^{\mathrm{T}} = -A$. Therefore,

$$\det A = \det A^{\mathrm{T}} = \det(-A) = (-1)^{2013} \det A = -\det A,$$

whence $\det A = 0$.
Answer: $\det(a_{ij}) = 0$.

2 Let a function $f : [a, b] \to \mathbb{R}$ be monotone, and F be one of its primitive functions. Show that $f \in C([a, b])$. Indeed, let $x_0 \in [a, b)$. By Lagrange's mean value theorem for each $0 < h < b - x_0$ there exists a real number $\theta_h \in (x_0, x_0 + h)$ such that $\frac{F(x_0+h)-F(x_0)}{h} = f(\theta_h)$. Since f is monotone, there exists the right-hand limit $f(x_0+)$. Thus $f(x_0) = F'(x_0) = \lim\limits_{h\to 0+} f(\theta_h) = f(x_0+)$. Similarly, for $x_0 \in (a, b]$ it holds $f(x_0) = f(x_0-)$. Therefore, $f \in C([a, b])$, whence by Cantor's theorem f is uniformly continuous on $[a, b]$.
Answer: yes, it is.

3 Consider a Cartesian coordinate system, in which the ellipse has a canonical equation $\frac{x^2}{a^2} + \frac{y^2}{b^2} = 1$, $a > b > 0$ (see Fig. 1). The foci F_1 and F_2 have coordinates $(\pm c, 0)$, where $c = \sqrt{a^2 - b^2}$. Take any point $A(x, y)$ of the ellipse that is not lying on the x-axis, and denote by I the incenter of the triangle F_1F_2A. Then I lies on the angle bisector AL of the triangle. Put $F_1A = u$ and $F_2A = v$, $u + v = 2a$. Let the point L have coordinates $(x_L, 0)$. By the angle bisector theorem $F_1L : F_2L = u : v$, therefore,

$$F_1L = x_L + c = F_1F_2 \cdot \frac{u}{u+v} = \frac{2cu}{u+v},$$

V. Brayman and A. Kukush, *Undergraduate Mathematics Competitions (1995–2016)*, Problem Books in Mathematics,
DOI 10.1007/978-3-319-58673-1_40

whence

$$x_L = \frac{c(u-v)}{u+v} = \frac{c(u^2-v^2)}{(u+v)^2} = \frac{c(u^2-v^2)}{4a^2}.$$

Next, $u^2 - v^2 = (x+c)^2 + y^2 - (x-c)^2 - y^2 = 4cx$, hence $x_L = \frac{c^2 x}{a^2}$.

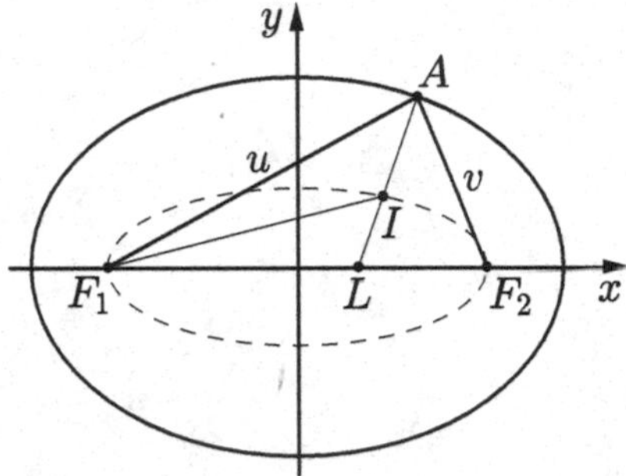

Fig. 1 The locus of incenters of triangles F_1F_2A, for which the vertex A lies on a given ellipse with the foci F_1 and F_2

Again by the angle bisector theorem the point $I(x_I, y_I)$ divides the segment AL in the ratio

$$AF_1 : F_1L = u : \frac{2cu}{u+v} = (u+v) : 2c = a : c,$$

whence

$$x_I = \frac{cx + ax_L}{a+c} = \frac{cx}{a}, \quad y_I = \frac{cy}{a+c}.$$

Thus, the point I belongs to an ellipse which can be obtained from the initial one after scaling along x-axes by a factor of $\frac{c}{a}$ and along y-axes by a factor of $\frac{c}{a+c}$. *Answer:* the locus is an ellipse, the major axis of which is F_1F_2, the points F_1 and F_2 omitted.

4 By the AM-GM inequality

$$\frac{1}{x_1^{x_2}} + \frac{1}{x_2^{x_3}} + \ldots + \frac{1}{x_n^{x_1}} \geq \frac{n}{\sqrt[n]{x_1^{x_2} x_2^{x_3} \ldots x_n^{x_1}}}.$$

By the weighted AM-GM inequality

$$\left(x_1^{x_2} x_2^{x_3} \ldots x_n^{x_1}\right)^{\frac{1}{x_1+\ldots+x_n}} \leq \frac{x_1x_2 + x_2x_3 + \ldots + x_nx_1}{x_1 + \ldots + x_n},$$

hence

$$\sqrt[n]{x_1^{x_2} x_2^{x_3} \ldots x_n^{x_1}} \leq \left(\frac{x_1x_2 + x_2x_3 + \ldots + x_nx_1}{x_1 + \ldots + x_n}\right)^{\frac{x_1+\ldots+x_n}{n}} \leq \frac{x_1x_2 + x_2x_3 + \ldots + x_nx_1}{x_1 + \ldots + x_n}$$

(indeed, the expression in parentheses does not exceed 1 and the exponent is not less than 1).

Thus,

$$\frac{1}{x_1^{x_2}} + \frac{1}{x_2^{x_3}} + \ldots + \frac{1}{x_n^{x_1}} \geq \frac{n(x_1 + \ldots + x_n)}{x_1x_2 + x_2x_3 + \ldots + x_nx_1} \geq$$

$$\geq \frac{n^2}{x_1x_2 + x_2x_3 + \ldots + x_nx_1} \geq \frac{n^2}{x_1^2 + \ldots + x_n^2}.$$

5 Let $n \geq 2$ and $m = [\sqrt{n}\,]$, where $[x]$ denotes the integer part of x. Each of m numbers $x_1, \ldots, x_m$ is a nonnegative integer and does not exceed n. Hence these numbers can be chosen in $(n+1)^m$ ways at most. Next, the sum of numbers $x_{m+1}, \ldots, x_n$ is less than or equal to m, otherwise

$$x_1 + 2x_2 + \ldots + nx_n \geq (m+1)(x_{m+1} + \ldots + x_n) \geq (m+1)^2 > n.$$

Thus, each choice of numbers $x_{m+1}, \ldots, x_n$ corresponds to a map from $\{1, \ldots, m\}$ to $\{m+1, \ldots, n, n+1\}$ which takes the value k at x_k points, $m+1 \leq k \leq n$, and the value $n+1$ at $m - (x_{m+1} + \ldots + x_n)$ points. It is clear that different choices of the numbers $x_{m+1}, \ldots, x_n$ correspond to different functions. Therefore, the number of ways to make a choice does not exceed $(n-m+1)^m \leq n^m$. Thus, a total number of solutions satisfies

$$p(n) \leq (n+1)^m n^m \leq (n+1)^{2m} \leq \exp\left(2\sqrt{n}\ln(n+1)\right) = O(e^{\varepsilon n}), \text{ as } n \to \infty,$$

for each $\varepsilon > 0$. Hence $p(n) = O(\alpha^n)$, as $n \to \infty$, for all $\alpha > 1$.

6 For $x = y = 0$, we have $f(0) = 0$. For $x = 0$, we get $f(-y) = f(y)$, for all $y \in \mathbb{R}$, i.e., f is an even function. Substitute $y = nx$ and get

$$f((n+1)x) = 2f(x) + 2f(nx) - f((n-1)x), \; n \in \mathbb{N},$$

whence it is easy to show by induction on $n \geq 1$ that $f(nx) = n^2 f(x)$, for all $x \in \mathbb{R}$, $n \in \mathbb{N}$. For every $x \in \mathbb{R}$ and $n, k \in \mathbb{N}$, it holds $f(nx) = k^2 f(\frac{n}{k}x) = n^2 f(x)$, whence $f(\frac{n}{k}x) = (\frac{n}{k})^2 f(x)$. Since f is even we get

$$f(rx) = r^2 f(x), \; x \in \mathbb{R}, \; r \in \mathbb{Q}.$$

For an arbitrary $x \neq 0$, consider a sequence $\{r_n\}$ of rational numbers that converges to $\frac{1}{x}$. Then $f(r_n x) = r_n^2 f(x) \to \frac{1}{x^2} f(x)$, as $n \to \infty$. But f is continuous at the point 1, hence $f(r_n x) \to f(1)$, as $n \to \infty$. Therefore, $\frac{1}{x^2} f(x) = f(1)$, that is $f(x) = x^2 f(1)$, for all $x \neq 0$. For $x = 0$, the latter equality still holds. Thus, $f(x) = ax^2$, $x \in \mathbb{R}$, where $a = f(1)$. It remains to check that these functions satisfy the conditions of the problem.

Answer: $f(x) = ax^2$, $x \in \mathbb{R}$, where $a \in \mathbb{R}$ is arbitrary.

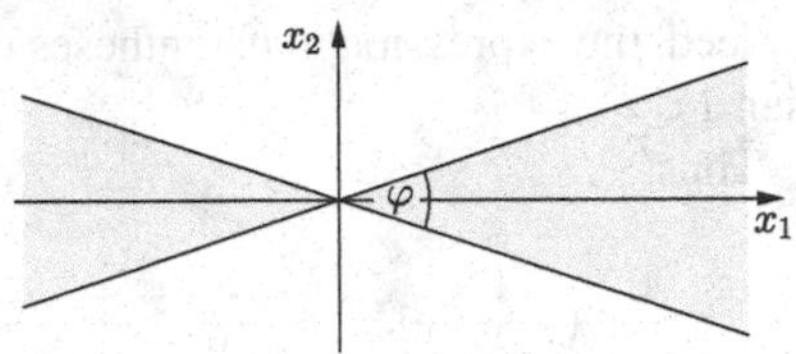

Fig. 2 The set of points (x_1, x_2) which satisfy the inequality $|x_2| \le \frac{1}{3}|x_1|$

7 Introduce a sequence $b_n = \sqrt{2(a_n^2 - 1)}$. It holds $b_1 = 0$ and $a_{n+1} = 3a_n + 2b_n$, $n \ge 1$. We claim that $b_{n+1} = 4a_n + 3b_n$, $n \ge 1$. Indeed, $a_n \ge 0$, $b_n \ge 0$, and

$$\begin{aligned} b_{n+1}^2 = 2a_{n+1}^2 - 2 &= 2(3a_n + 2b_n)^2 - 2a_n^2 + b_n^2 = \\ &= 16a_n^2 + 9b_n^2 + 24a_nb_n = (4a_n + 3b_n)^2. \end{aligned}$$

Therefore, $\begin{pmatrix} a_1 \\ b_1 \end{pmatrix} = \begin{pmatrix} 1 \\ 0 \end{pmatrix}$ and $\begin{pmatrix} a_{n+1} \\ b_{n+1} \end{pmatrix} = \begin{pmatrix} 3 & 2 \\ 4 & 3 \end{pmatrix} \begin{pmatrix} a_n \\ b_n \end{pmatrix}$, $n \ge 1$, whence all the numbers a_n and b_n, $n \ge 2$, are positive integers.

Fix an arbitrary $n \ge 1$. Introduce matrices $A = \begin{pmatrix} 3 & 2 \\ 4 & 3 \end{pmatrix}$ and $B = A^n = \begin{pmatrix} p & q \\ r & s \end{pmatrix}$. Then

$$\begin{pmatrix} a_{n+1} \\ b_{n+1} \end{pmatrix} = A^n \begin{pmatrix} a_1 \\ b_1 \end{pmatrix} = B \begin{pmatrix} 1 \\ 0 \end{pmatrix}, \quad \begin{pmatrix} a_{2n+1} \\ b_{2n+1} \end{pmatrix} = A^{2n} \begin{pmatrix} a_1 \\ b_1 \end{pmatrix} = B^2 \begin{pmatrix} 1 \\ 0 \end{pmatrix},$$

whence $a_{n+1} = p$, $a_{2n+1} = p^2 + qr$. It remains to notice that

$$\det B = ps - qr = (\det A)^n = 1.$$

Therefore, p and qr are relatively prime, hence a_{n+1} and a_{2n+1} are relatively prime as well.

Remark 1 One can show that $a_{n+2} = 6a_{n+1} - a_n$, $n \ge 1$. The characteristic equation $\lambda^2 = 6\lambda - 1$ has roots $3 \pm \sqrt{8}$, hence $a_n = \alpha\left(3 - \sqrt{8}\right)^n + \beta\left(3 + \sqrt{8}\right)^n$, where α and β can be found from the initial conditions $a_1 = 1$, $a_2 = 3$. Then

$$a_n = \tfrac{1}{2}\left(\left(3 - \sqrt{8}\right)^{n-1} + \left(3 + \sqrt{8}\right)^{n-1}\right), \quad n \ge 1.$$

Now, it is easy to check that $a_{2n+1} = 2a_{n+1}^2 - 1$, therefore, a_{n+1} and a_{2n+1} are relatively prime.

8 Put $A = \begin{pmatrix} \frac{1}{\sqrt{2}} & 0 \\ 0 & \sqrt{3} \end{pmatrix}$, $\det A = \sqrt{\frac{3}{2}} > 1$. Find for which vectors $x = \begin{pmatrix} x_1 \\ x_2 \end{pmatrix} \in \mathbb{R}^2$ the inequality $\|Ax\| \le \frac{\sqrt{3}}{2}\|x\|$ holds. This inequality is equivalent to

$$\frac{1}{2}x_1^2 + 3x_2^2 \le \frac{3}{4}(x_1^2 + x_2^2), \text{ or } |x_2| \le \frac{1}{3}|x_1|.$$

The set of points (x_1, x_2) which satisfy the latter inequality is shaded on Fig. 2. Since $\tan\frac{\varphi}{2} = \frac{1}{3}$ then $\tan\varphi = \frac{3}{4}$, whence $\varphi > \frac{\pi}{6}$.

Put $B = (1+\varepsilon)\begin{pmatrix}\cos\varphi & -\sin\varphi \\ \sin\varphi & \cos\varphi\end{pmatrix}$, where $\varepsilon > 0$ is such that $(1+\varepsilon)^5 \cdot \frac{\sqrt{3}}{2} \le 1$. We have $\det B = (1+\varepsilon)^2 > 1$. Show that the matrices A and B have required properties.

For a fixed $u_0 \in \mathbb{R}^2$, introduce a vector sequence u_i, $i \ge 1$, as follows: $u_i = M_i u_{i-1}$, where $M_i = A$ if u_{i-1} belongs to the shaded set on Fig. 2, and $M_i = B$ otherwise. Put $i_0 = 0$, and let $1 \le i_1 < i_2 < \ldots < i_k < \ldots$ be the sequence of indices for which $M_i = A$. To obtain a vector Bx one should rotate a vector x by the angle φ counterclockwise and increase the length of the resulting vector in $1+\varepsilon$ times. Since $\varphi > \frac{\pi}{6}$, then for each $x \in \mathbb{R}^2$ at least one of the vectors $x, Bx, B^2x, \ldots, B^5x$ belongs to the shaded set on Fig. 2. Thus, $i_1 \le 5$, $i_2 - i_1 \le 5$, $i_3 - i_2 \le 5, \ldots$ Therefore, the inequality $(1+\varepsilon)^5 \cdot \frac{\sqrt{3}}{2} \le 1$ implies $\|u_0\| \ge \|u_{i_1}\| \ge \|u_{i_2}\| \ge \|u_{i_3}\| \ge \ldots$ Then for $i_k < i \le i_{k+1}$, $k \ge 0$, it holds $\|u_i\| \le (1+\varepsilon)^5\|u_{i_k}\| \le (1+\varepsilon)^5\|u_0\|$.
Answer: yes, there exist.

9 Let $1 \le k \le n-1$. Rewrite the inequality in question:

$$\frac{1}{n-k}\sum_{j=k}^{n}\frac{n-j}{j} \le 1, \quad \frac{1}{n-k}\sum_{j=k}^{n-1}\left(\frac{n}{j}-1\right) \le 1, \quad \frac{1}{n-k}\sum_{j=k}^{n-1}\frac{n}{j} \le 2,$$

$$\sum_{j=k}^{n-1}\frac{1}{j} \le \frac{2(n-k)}{n} = 2 - \frac{2k}{n}.$$

Since $n \ge 3$ by the condition of the problem, the inequality holds for $k = n-1$ and fails for $k = 1$, because $\frac{1}{n-1} \le \frac{2}{n}$ and $1 + \frac{1}{2} + \ldots + \frac{1}{n-1} + \frac{2}{n} > 1 + \frac{n}{n} = 2$. Hence there exists the required number k^*, and moreover $2 \le k^* \le n-1$. Also notice that for each fixed k the inequality fails for n large enough, hence $k^* \to \infty$, as $n \to \infty$.

The number k^* satisfies

$$2 - \frac{2k^*}{n} \ge \sum_{j=k^*}^{n-1}\frac{1}{j}, \quad 2 - \frac{2k^*-2}{n} < \sum_{j=k^*-1}^{n-1}\frac{1}{j}, \tag{$*$}$$

hence $\frac{1}{k^*-1} > \frac{2}{n}$, that is $k^* < \frac{n}{2} + 1$.

Show that a sequence $\left\{\frac{k^*(n)}{n}\right\}$ has a unique limit point. Consider an arbitrary subsequence of numbers $\{n'\}$, for which $\frac{k^*(n')}{n'} \to u \in \left[0, \frac{1}{2}\right]$, as $n' \to \infty$. Because $k^*(n') \to \infty$, as $n' \to \infty$, the inequalities $(*)$ imply that

$$\sum_{j=k^*(n')}^{n'-1} \frac{1}{j} \to 2 - 2u, \text{ as } n' \to \infty.$$

On the other hand, since

$$\left| \sum_{j=k}^{n-1} \frac{1}{j} - \sum_{j=k}^{n-1} \int_j^{j+1} \frac{dt}{t} \right| \le \sum_{j=k}^{n-1} \left(\frac{1}{j} - \frac{1}{j+1} \right) < \frac{1}{k} \text{ and } \sum_{j=k}^{n-1} \int_j^{j+1} \frac{dt}{t} = \int_k^n \frac{dt}{t} = \ln \frac{n}{k},$$

it follows that

$$\lim_{n' \to \infty} \sum_{j=k^*(n')}^{n'-1} \frac{1}{j} = \lim_{n' \to \infty} \ln \frac{n'}{k^*(n')} = -\ln u.$$

Thus, $2 - 2u = -\ln u$. The function $f(x) = 2 - 2x + \ln x$ is increasing on $(0, \frac{1}{2}]$, $f(0+) = -\infty$, and $f(\frac{1}{2}) = 1 - \ln 2 > 0$. Hence this function has a unique zero x_0 on the interval $(0, \frac{1}{2})$. We have $f(u) = 0$, so $u = x_0$ is the only possible limit point of the sequence $\left\{ \frac{k^*(n)}{n} \right\}$, hence $\frac{k^*(n)}{n} \to x_0$, as $n \to \infty$.

10 Fix $x_1, x_2 \in X$ and $y_1, y_2 \in Y$ such that $x_1 \neq x_2$, $y_1 \neq y_2$. Introduce a continuous function

$$f(x, y) = \rho(x, x_1)\sigma(y, y_1), \; x \in X, y \in Y.$$

Suppose that $f(x, y) = g(x) + h(y)$, for all $x \in X$ and $y \in Y$. Substitute $y = y_1$ and get $f(x, y_1) \equiv 0$, whence $g(x) \equiv -h(y_1)$. In a similar way substitute $x = x_1$ and get $h(y) \equiv -g(x_1)$. Hence $f(x, y) \equiv -h(y_1) - g(x_1) = \text{const}$. But $f(x_1, y_1) = 0$ and $f(x_2, y_2) = \rho(x_2, x_1)\sigma(y_2, y_1) > 0$, a contradiction.
Answer: yes, always.

11 If $\xi = 0$ or $\eta = 0$ then $u(\xi + \eta) - u(\xi) - u(\eta) + u(0) = 0$. Since by the condition of the problem it holds $\mathsf{P}(\xi = 0 \text{ or } \eta = 0) = 1$, we have

$$\mathsf{E}\,(u(\xi + \eta) - u(\xi) - u(\eta) + u(0)) = 0.$$

12 The general solution to the equation has a form

$$x(t) = ce^{-at} + e^{-at} \int_0^t e^{au} f(u)\, du, \; t \in \mathbb{R}, \qquad (*)$$

where c is an arbitrary constant. Denote $f_\infty = \lim_{t \to +\infty} f(t)$.

(a) If $a > 0$ then the first summand on the right-hand side of $(*)$ tends to zero, as $t \to +\infty$, and one can apply L'Hospital's Rule to the second summand, because $e^{at} \to +\infty$, as $t \to +\infty$:

$$\lim_{t\to+\infty}\frac{1}{e^{at}}\int_0^t e^{au}f(u)\,du=\lim_{t\to+\infty}\frac{\frac{d}{dt}\int_0^t e^{au}f(u)\,du}{\frac{d}{dt}e^{at}}=\lim_{t\to+\infty}\frac{e^{at}f(t)}{ae^{at}}=\frac{f_\infty}{a}.$$

Thus, the general solution (∗) tends to $\frac{f_\infty}{a}$.
Answer: for $a>0$, the limit equals $\frac{1}{a}\lim_{t\to+\infty}f(t)$.

(b) Let $a<0$. Notice that the continuous function f is bounded on $[0,+\infty)$, hence the improper integral $\int_0^{+\infty}e^{au}f(u)\,du$ converges absolutely. Rewrite the solution in a form

$$x(t)=\left(c+\int_0^\infty e^{au}f(u)\,du\right)e^{-at}-\frac{1}{e^{at}}\int_t^{+\infty}e^{au}f(u)\,du,\ t\in\mathbb{R}.$$

As $t\to+\infty$, we have an indeterminate form $0/0$ in the second summand. By L'Hospital's Rule we get

$$\lim_{t\to+\infty}\frac{1}{e^{at}}\int_t^{+\infty}e^{au}f(u)\,du=\lim_{t\to+\infty}\frac{-e^{at}f(t)}{ae^{at}}=\frac{f_\infty}{-a}.$$

Therefore, the solution has a finite limit, as $t\to+\infty$, only for $c=-\int_0^{+\infty}e^{au}f(u)\,du$, and in this case the limit equals $\frac{f_\infty}{a}$.

13 Because $A^2=0$, the matrix A has a single eigenvalue 0 of multiplicity 3. Moreover, the Jordan form of A does not equal to the cell $J_3(0)$, i.e., the cells in the Jordan form of A are $J_2(0)$ or $J_1(0)$. Hence $\mathrm{rk}(A)\le 1$. In a similar way $\mathrm{rk}(B)\le 1$. Therefore, $\mathrm{rk}(A+B)\le 2$, and 0 is an eigenvalue of the matrix $A+B$. Moreover $\mathrm{tr}(A+B)=0$, i.e., the sum of all the eigenvalues of $A+B$ (repeated according to their multiplicities) equals 0. Hence the set of eigenvalues of $A+B$ has a form $\{0,\lambda,-\lambda\}$ for some $\lambda\in\mathbb{C}$. Every such set is indeed a collection of eigenvalues of $A+B$ for the matrices

$$A=\begin{pmatrix}0&\lambda&0\\0&0&0\\0&0&0\end{pmatrix},\quad B=\begin{pmatrix}0&0&0\\\lambda&0&0\\0&0&0\end{pmatrix}.$$

Answer: $\{0,\lambda,-\lambda\}$, where $\lambda\in\mathbb{C}$ is arbitrary.

14 Show that $g_n\xrightarrow{\lambda}0$, as $n\to\infty$. Suppose that it is false. Then there exists a sequence $n_k\to\infty$ and $\varepsilon>0$ such that $\lambda(\{g_{n_k}\ge\varepsilon\})\ge\varepsilon,\ k\ge 1$. The function $f(x)=\frac{x^2}{1+x}$ is increasing on $[0,+\infty)$, because $f'(x)=\frac{x^2+2x}{(1+x)^2}>0$. Hence

$$\int_X \frac{g_{n_k}^2}{1+g_{n_k}} d\lambda \geq \varepsilon \cdot \frac{\varepsilon^2}{1+\varepsilon} \not\to 0, \text{ as } k \to \infty,$$

a contradiction.

Then $\int\limits_X g_n I_{\{g_n<1\}} d\lambda \to 0$, as $n \to \infty$, by Lebesgue dominated convergence theorem (with the majorant 1), and $\int\limits_X g_n I_{\{g_n \geq 1\}} d\lambda \to 0$, as $n \to \infty$, because $g_n \leq \frac{2g_n^2}{1+g_n}$ on the set $\{x \mid g_n(x) \geq 1\}$.

15 Fix $a \in [0, 1]$ and an orthobasis $e_0, e_1, \ldots$ in l_2. Then the vectors

$$v_k = \sqrt{a}e_0 + \sqrt{1-a}e_k, \ k \geq 1,$$

satisfy the condition of the problem.

Conversely, assume that vectors v_k, $k \geq 1$, satisfy the condition of the problem. Prove that $a \in [0, 1]$. Let $a = x + iy$, where $x, y \in \mathbb{R}$. By the Cauchy–Schwarz inequality it holds

$$|(v_1 + \ldots + v_n, v_{n+1} + \ldots + v_{2n})| \leq \|v_1 + \ldots + v_n\| \cdot \|v_{n+1} + \ldots + v_{2n}\|.$$

Notice that

$$\|v_1 + \ldots + v_n\|^2 = \sum_{k=1}^{n} \|v_k\|^2 + 2\sum_{k<l} \operatorname{Re}(v_k, v_l) = n + n(n-1)x,$$

and in a similar way $\|v_{n+1} + \ldots + v_{2n}\|^2 = n + n(n-1)x$. Hence

$$n^2|x + iy| \leq n + n(n-1)x, \ \text{ i.e., } \sqrt{x^2+y^2} \leq \tfrac{1}{n} + \tfrac{n-1}{n}x.$$

As $n \to \infty$, we get $\sqrt{x^2+y^2} \leq x$. Therefore, $x \geq 0$ and $y = 0$. Thus, $a = x$. Since $|(v_1, v_2)| \leq 1$ we have $x \leq 1$.
Answer: $a \in [0, 1]$.

16 We perform the proof in several steps.
(1) Similarly to the solution of Problem 6, we get that $f(0) = 0$ and f is even.
(2) Show that for all $x, y \in \mathbb{R}$ and $n \in \mathbb{Z}$ it holds

$$f(nx + y) = n^2 f(x) + n\Delta + f(y), \tag{$*$}$$

where $\Delta = \Delta(x, y) = f(x + y) - f(x) - f(y)$. For $n = 0$ and $n = 1$ the equality is evident. Next, $f(kx + y + x) + f(kx + y - x) = 2f(kx + y) + 2f(x)$. Thus, if $(*)$ holds for $n = k - 1$ and $n = k$, then

$$\begin{aligned} f((k+1)x+y) &= 2f(kx+y)+2f(x)-f((k-1)x+y) = \\ &= \left(2k^2+2-(k-1)^2\right)f(x)+(2k-(k-1))\,\Delta+(2-1)f(y) = \\ &= (k+1)^2 f(x)+(k+1)\Delta+f(y), \end{aligned}$$

i.e., $(*)$ holds for $n=k+1$ as well. Similarly if $(*)$ holds for $n=k$ and $n=k+1$, then $(*)$ holds for $n=k-1$ as well. Therefore, by the principle of mathematical induction $(*)$ holds for all $n \in \mathbb{Z}$.

(3) We substitute $y=0$ in $(*)$ and obtain $f(nx)=n^2 f(x)$, for all $x \in \mathbb{R}$ and $n \in \mathbb{N}$. Hence for all $x, y \in \mathbb{R}$ and $n \in \mathbb{Z}$, it holds

$$\begin{aligned} \Delta(nx, y) &= f(nx+y)-f(nx)-f(y) = \\ &= n^2 f(x)+n\Delta(x,y)+f(y)-n^2 f(x)-f(y)=n\Delta(x,y). \end{aligned}$$

(4) Steps (2) and (3) imply that

$$f(qx+y)=q^2 f(x)+q\Delta(x,y)+f(y)$$

for every $x, y \in \mathbb{R}$ and $q \in \mathbb{Q}$. Indeed, if $q=\frac{m}{n}$, $m \in \mathbb{Z}$, $n \in \mathbb{N}$, then

$$f\left(\tfrac{m}{n}x+y\right)=m^2 f\left(\tfrac{x}{n}\right)+m\Delta\left(\tfrac{x}{n}, y\right)+f(y)=\tfrac{m^2}{n^2}f(x)+\tfrac{m}{n}\Delta(x,y)+f(y).$$

(5) To prove the statement of the problem it suffices to show that

$$\frac{f(x+y)}{x+y}=\frac{f(x)}{x}+\frac{f(y)}{y}, \quad \text{for all } x \neq 0,\ y \neq 0,\ x+y \neq 0. \qquad (**)$$

To this purpose take a sequence $\{q_n,\ n \geq 1\} \subset \mathbb{Q}$ such that $q_n \to -\frac{y}{x}$, as $n \to \infty$. We have

$$f(q_n x+y)=q_n^2 f(x)+q_n\left(f(x+y)-f(x)-f(y)\right)+f(y).$$

Pass to the limit in the equality as $n \to \infty$. Since f is continuous at zero, we get

$$0=f(0)=\frac{y^2}{x^2}f(x)-\frac{y}{x}\left(f(x+y)-f(x)-f(y)\right)+f(y),$$

or equivalently,

$$\frac{(x+y)y}{x^2}f(x)+\frac{x+y}{x}f(y)=\frac{y}{x}f(x+y),$$

which implies $(**)$.

Remark 1 The conditions of the problem imply that $f(x) \equiv ax^2$, $x \in \mathbb{R}$, for some $a \in \mathbb{R}$. Indeed, consider the function φ from the statement of the problem. The function satisfies $\varphi(x + y) \equiv \varphi(x) + \varphi(y)$. Then either $\varphi(x) = ax$ for some fixed $a \in \mathbb{R}$, or the graph of φ is dense in $\mathbb{R}^2$. If $\varphi(x) = ax$ then $f(x) = ax^2$. Show that the graph of φ cannot be dense in $\mathbb{R}^2$. The function f is continuous at zero, hence there exists $\delta > 0$ such that $|f(x)| \leq 1$ for all $x \in (-\delta, \delta)$. Therefore, $|\varphi(x)| \leq \frac{1}{|x|}$ for all $x \in (0, \delta)$, and the graph of φ is not dense in $\mathbb{R}^2$.

2013

1 Substitute $x = 1$ and get $f(2) = 1$. Because $f(1) = 2$ and $f(2) = 1$, by the intermediate value theorem for each $y \in [1, 2]$ there exists $z \in [1, 2]$ such that $f(z) = y$. Substitute $x = z$ and get $f(y) \cdot y = 2$, i.e. $f(y) = \frac{2}{y}$, $y \in [1, 2]$. It is easy to check that this function satisfies the conditions of the problem.
Answer: $f(x) = \frac{2}{x}$, $x \in [1, 2]$.

2 If there exists a finite ring with required properties, then the squares of its elements are pairwise distinct (indeed, otherwise there exists an element which is not a square of any element of the ring). But by the condition of the problem there exists some $y \neq 0$ such that $y^2 = 0$, and $0^2 = 0$ (here 0 is the zero element of the ring), a contradiction.
Answer: no, there is no such a ring.

3 Let the angles of triangle ABC be such that $\tan A + \tan C = 2 \tan B$. Then

$$\tan B = -\tan(A + C) = \frac{\tan A + \tan C}{\tan A \tan C - 1} = \frac{2 \tan B}{\tan A \tan C - 1},$$

hence $\tan A \tan C = 3$. Thus, both numbers $\tan A$ and $\tan C$ are positive, therefore, $\tan B$ is positive as well, and the triangle is acute. Since the function $y = \tan x$ is increasing on $\left[0, \frac{\pi}{2}\right)$, we have that B is a middle angle of the triangle. The Law of Sines implies that sines of the angles form an arithmetic progression as well. The function $y = \sin x$ is increasing on $\left[0, \frac{\pi}{2}\right)$, hence $\sin B$ is the middle term of the progression, i.e., $\sin A + \sin C = 2 \sin B$. Next, the function $y = \tan x$ is convex on $\left[0, \frac{\pi}{2}\right)$, and the function $y = \sin x$ is concave on $\left[0, \frac{\pi}{2}\right)$. Therefore,

$$\sin B = \tfrac{\sin A + \sin C}{2} \le \sin \tfrac{A+C}{2} \quad \text{and} \quad \tan B = \tfrac{\tan A + \tan C}{2} \ge \tan \tfrac{A+C}{2},$$

V. Brayman and A. Kukush, *Undergraduate Mathematics Competitions (1995–2016)*, Problem Books in Mathematics,
DOI 10.1007/978-3-319-58673-1_41

whence $B \le \frac{A+C}{2}$ and $B \ge \frac{A+C}{2}$. Thus, $B = \frac{A+C}{2} = \frac{\pi - B}{2}$, that is $B = \frac{\pi}{3}$. From $\tan A + \tan C = 2\sqrt{3}$ and $\tan A \tan C = 3$ we derive that $\tan A = \tan C = \sqrt{3}$, and finally $A = B = C = \frac{\pi}{3}$.
Answer: each angle equals $\frac{\pi}{3}$, i.e., the triangle is equilateral.

4 Introduce a function

$$f(x) = \sqrt{x_1 + \sqrt{x_2 + \sqrt{\ldots + \sqrt{x_n + x}}}}, \quad x \ge 0.$$

By Lagrange's mean value theorem there exists $\theta \in (0, c)$ for which $f(c) - f(0) = cf'(\theta)$. We have

$$f'(\theta) = \frac{1}{2\sqrt{x_1 + \sqrt{x_2 + \sqrt{\ldots + \sqrt{x_n + \theta}}}}} \times \\ \times \frac{1}{2\sqrt{x_2 + \sqrt{\ldots + \sqrt{x_n + \theta}}}} \cdot \ldots \cdot \frac{1}{2\sqrt{x_n + \theta}},$$

hence $f'(\theta) < \frac{1}{2^n \sqrt{x_1 \cdot \ldots \cdot x_n}}$ and $f(c) < f(0) + \frac{c}{2^n \sqrt{x_1 \cdot \ldots \cdot x_n}}$.

5 Show that at least one of the matrices A and B is nonsingular. Indeed, if the matrix A is singular, by elementary transformations of rows one can nullify its first row, hence there exists some nonsingular matrix U such that the first row of the matrix UA is zero. Similarly if the matrix B is singular, then there exists some nonsingular matrix V such that the first row of the matrix BV is zero. Notice the following: if X, Y is a solution of the equation $AX + YB = U^{-1}CV^{-1}$, then $UA(XV) + (UY)BV = C$, i.e., the equation $UAX + YBV = C$ has a solution for every matrix C as well. But for arbitrary matrices X and Y, the first row of UAX and the first column of YBV are zero. Hence if the entry in the first row and the first column of C is nonzero, the equation $UAX + YBV = C$ has no solution, a contradiction.

If the matrix A is nonsingular, then the equation $A^{2013}X + YB^{2013} = C$ has a solution $X = A^{-2013}C$ and $Y = 0$, and if B is nonsingular, then one can put $X = 0$ and $Y = CB^{-2013}$, respectively.

6 It suffices to prove that the set $A = \{x \in \mathbb{R} \mid f(x) + g(x) < 0\}$ is at most countable. Fix some $\delta > 0$, and show that the set $A_\delta = \{x \in \mathbb{R} \mid f(x) + g(x) < -\delta\}$ is at most countable. Take any $x, y \in A_\delta$. Then $f(x) + g(x) < -\delta$, $f(y) + g(y) < -\delta$, and at least one of the inequalities $f(x) + g(y) > 0$ or $f(y) + g(x) > 0$ holds.

If $f(x) + g(y) > 0$ then $f(x) - f(y) > -(f(y) + g(y)) > \delta$, and if $f(y) + g(x) > 0$ then $f(x) - f(y) < f(x) + g(x) < -\delta$. Thus, for arbitrary distinct points $x, y \in A_\delta$, it holds $|f(x) - f(y)| > \delta$. Hence each of the sets $B_{k,\delta} = \{x \in \mathbb{R} \mid f(x) \in [k\delta, k\delta + \delta)\}$, $k \in \mathbb{Z}$, contains no more than one point from A_δ. Since $\bigcup_{k \in \mathbb{Z}} B_{k,\delta} = \mathbb{R}$, the set A_δ is at most countable for each $\delta > 0$. Hence the set $A = \bigcup_{n\geq 1} A_{1/n}$ is at most countable as well.

7 Fix a positive integer N, and estimate the number of positive integers that are less than N and cannot be represented as a sum of two *good* numbers. Squares of positive integers have remainders 0 or 1 after division by 4, hence no number of the form $4j + 3$ is a sum of two squares.

Let $n^k + m^l < N$, where $k \geq 3$ and $l \geq 2$. We have $n \leq N^{1/3}$, $m \leq N^{1/2}$, and it holds $k, l \leq \log_2 N$ for $n, m \geq 2$ (one can assume that $k = 3$ for $n = 1$ and $l = 2$ for $m = 1$). Thus, there exist at most $N^{5/6}(\log_2 N)^2$ numbers which are less than N and can be represented as a sum of two *good* non-square numbers. Consider the numbers of the form $4j + 3$ which are less than N. There are at least $\left[\frac{N}{4}\right] - N^{5/6}(\log_2 N)^2$ numbers among them which cannot be represented as a sum of two *good* ones, and $\left[\frac{N}{4}\right] - N^{5/6}(\log_2 N)^2 \to \infty$, as $N \to \infty$.
Answer: it is infinite.

8 We say that a matrix $M = \left(\begin{smallmatrix} a & b \\ c & d \end{smallmatrix}\right)$ is of type I if it holds

$$|a| > |c|, \quad |b| > |d|, \quad ac \geq 0, \quad bd \geq 0,$$

and of type II if it holds

$$|a| < |c|, \quad |b| < |d|, \quad ac \leq 0, \quad bd \leq 0.$$

In particular, the matrix $A = \left(\begin{smallmatrix} 1 & 2 \\ 0 & 1 \end{smallmatrix}\right)$ is of type I, and the matrix $B = \left(\begin{smallmatrix} 1 & 0 \\ -2 & 1 \end{smallmatrix}\right)$ is of type II.

Consider the matrix $AM = \left(\begin{smallmatrix} a+2c & b+2d \\ c & d \end{smallmatrix}\right)$. If the matrix M is of type I then

$$\begin{gathered} |a + 2c| = |a| + 2|c| > |c|, \quad |b + 2d| = |b| + 2|d| > |d|, \\ (a + 2c)c = ac + 2c^2 \geq 0, \quad (b + 2d)d = bd + d^2 \geq 0, \end{gathered}$$

and if the matrix M is of type II then

$$\begin{gathered} |a + 2c| = 2|c| - |a| > |c|, \quad |b + 2d| = 2|d| - |b| > |d|, \\ (a + 2c)c = 2c^2 - (-ac) \geq 0, \quad (b + 2d)d = d^2 - (-bd) \geq 0, \end{gathered}$$

i.e. in both cases AM is of type I. In a similar way one can verify that if M is of type either I or II, then the matrix $BM = \begin{pmatrix} a & b \\ -2a + c & -2b + d \end{pmatrix}$ is of type II. Hence every product $X_1 X_2 \ldots X_n$, where each multiplier X_i equals either A or B, is of type I in case $X_1 = A$ and of type II in case $X_1 = B$. Therefore, the product cannot be equal to the identity matrix, which is neither of type I nor of type II.
Answer: no, it can not.

9 For all $x \in \mathbb{R}$, it holds

$$\sum_{n=0}^{\infty} \frac{\cos\left(\frac{\pi n}{2} - x\right)}{n!} x^n = \cos x \sum_{n=0}^{\infty} \frac{\cos \frac{\pi n}{2}}{n!} x^n + \sin x \sum_{n=0}^{\infty} \frac{\sin \frac{\pi n}{2}}{n!} x^n =$$

$$= \cos x \sum_{k=0}^{\infty} \frac{(-1)^k}{(2k)!} x^{2k} + \sin x \sum_{k=0}^{\infty} \frac{(-1)^k}{(2k+1)!} x^{2k+1} = \cos^2 x + \sin^2 x = 1.$$

In particular for $x = \frac{1}{2}$ we obtain the sum of the series in question.
Answer: 1.

10 Suppose that there exists a finite ring with required properties. Then the squares of its elements are pairwise distinct (otherwise there exists an element which cannot be represented as the square of an element of the ring). Fix an arbitrary nonzero element x of the ring. Consider a sequence $x, x^2, x^4, x^8, \ldots$ Since the ring is finite, some terms of the sequence are equal. Let $x^{2^{k+n}} = x^{2^k}$. Because the squares of all the elements of the ring are distinct, we get $x^{2^n} = x$. Denote $a = x^{2^n - 1} \neq 0$. Then

$$a^2 = (x^{2^n-1})^2 = x^{2^n} \cdot x^{2^n-2} = x \cdot x^{2^n-2} = a,$$

but by the condition of the problem there exists an element $b \neq a$ such that $b^2 = a$ as well, a contradiction.s
Answer: no, there is no such a ring.

12 It was shown in the solution of Problem 5 that at least one of the matrices A and B is nonsingular. Hence one of the numbers $k_0(A)$ and $k_0(B)$ equals 0, and the other is at most n.

13 Let

$$C_{xy} = \operatorname*{ess\,sup}_{x\in\mathbb{R}} \left(\operatorname*{ess\,sup}_{y\in\mathbb{R}} f(x, y) \right), \quad C_{yx} = \operatorname*{ess\,sup}_{y\in\mathbb{R}} \left(\operatorname*{ess\,sup}_{x\in\mathbb{R}} f(x, y) \right).$$

Denote by λ_1 the Lebesgue measure on $\mathbb{R}$ and by λ_2 the Lebesgue measure on $\mathbb{R}^2$. We claim that $C_{xy} = C_{yx} = C$, where $C = \operatorname*{ess\,sup}_{(x,y)\in\mathbb{R}^2} f(x, y)$ is essential supremum with respect to the measure λ_2.

For a fixed $a \in \mathbb{R}$, introduce sets

$$A_a = \{(x, y) \mid f(x, y) > a\}, \quad A_{a,x} = \{y \mid f(x, y) > a\}, \ x \in \mathbb{R}.$$

Then

$$\lambda_2(A_a) = \int_{\mathbb{R}} \lambda_1(A_{a,x}) d\lambda_1(x).$$

If $C < \infty$ then for $a = C$ it holds $\lambda_2(A_C) = 0$, whence $\lambda_1(A_{C,x}) = 0$ for almost all $x \in \mathbb{R}$. Then $\operatorname{ess\,sup}_{y\in\mathbb{R}} f(x, y) \le C$ for almost all $x \in \mathbb{R}$, hence $C_{xy} \le C$. It is evident that for $C = \infty$, the inequality $C_{xy} \le C$ holds as well.

For every $a < C$, we have $\lambda_2(A_C) > 0$. Therefore, there exists a set $B \subset \mathbb{R}$ such that $\lambda_1(B) > 0$ and $\lambda_1(A_{a,x}) > 0$, $x \in B$. Then $\operatorname{ess\,sup}_{y\in\mathbb{R}} f(x, y) > a$, $x \in B$, whence $C_{xy} > a$. Because $a < C$ is arbitrary, it holds $C_{xy} \ge C$.

The inequalities $C_{xy} \le C$ and $C_{xy} \ge C$ imply that $C_{xy} = C$. Similarly $C_{yx} = C$, hence $C_{xy} = C_{yx}$.

Answer: yes, it is always true.

14 For an irreducible polynomial p and a rational function φ, put $\operatorname{ord}_p(\varphi) = n \in \mathbb{Z}$ if $\varphi = p^n \cdot \frac{q}{r}$, where q and r are polynomials which are not divisible by p. Then

$$\varphi' = np^{n-1}p' \cdot \frac{q}{r} + p^n \cdot \frac{q'r - qr'}{r^2} = p^{n-1} \cdot \frac{np'qr + p(q'r - qr')}{r^2},$$

$$\frac{\varphi'}{\varphi} = \frac{np'qr + p(q'r - qr')}{pqr}.$$

Hence for $n \ne 0$, it holds $\operatorname{ord}_p(\varphi') = n - 1$ and $\operatorname{ord}_p(\frac{\varphi'}{\varphi}) = -1$, and for $n = 0$, we have $\operatorname{ord}_p(\varphi') \ge 0$.

Let φ and ψ be nonconstant rational functions, for which $\psi' = \frac{\varphi'}{\varphi}$. Consider an irreducible polynomial p, which is a divisor of numerator or denominator of the function φ. Then $\operatorname{ord}_p(\varphi) \ne 0$, hence $\operatorname{ord}_p(\psi') = \operatorname{ord}_p(\frac{\varphi'}{\varphi}) = -1$. But if $\operatorname{ord}_p(\psi) = k \ne 0$, then $\operatorname{ord}_p(\psi') = k - 1 \ne -1$, and if $\operatorname{ord}_p(\psi) = 0$, then $\operatorname{ord}_p(\psi) \ge 0$. In both cases we get a contradiction with $\operatorname{ord}_p(\psi') = -1$.

Answer: no, they do not exist.

15 For a finite measure μ on Borel sigma-algebra $\mathscr{B}(\mathbb{R}^2)$, denote by $\operatorname{supp}\mu$ the support of μ, that is the minimal closed set F for which $\mu(F) = \mu(\mathbb{R}^2)$.

Rewrite the statement of the problem as follows: if $\operatorname{supp}\mathsf{P}$ does not belong to any straight line, then there exists a bounded Borel measurable set A such that the support of the restricted measure $\mathsf{P}_A(B) = \mathsf{P}(A \cap B)$, $B \in \mathscr{B}(\mathbb{R}^2)$, does not belong to any straight line as well.

By the condition of the problem the support of P contains at least three noncollinear points u, v, w. It is not difficult to verify that $\operatorname{supp}\mathsf{P}_A = \operatorname{supp}\mathsf{P} \cap A$ for an arbitrary closed set A. Let A be a closed disc with center in the origin and radius $R = \max\{\|u\|, \|v\|, \|w\|\}$. Then the support of the measure P_A contains all three points u, v, w hence it does not belong to any straight line.

Answer: yes, it is.

16 Show that partial sums of the series $\sum_{n=1}^{\infty}(-1)^{J_n}a_n$ form a Cauchy sequence in $L_2(\Omega, \mathscr{F}, \mathsf{P})$. This will imply that the series converges in $L_2(\Omega, \mathscr{F}, \mathsf{P})$, hence in probability as well.

Denote $T_{k,k+m} = \sum_{n=k}^{k+m} (-1)^{S_n} a_n$. It holds

$$\mathsf{E}T_{k,k+m}^2 = \sum_{i,j=k}^{k+m} a_i a_j \mathsf{E}(-1)^{S_i+S_j}.$$

Next, $\mathsf{E}(-1)^{S_i+S_j} = \mathsf{E}(-1)^{S_i-S_j} = \mathsf{E}(-1)^{S_{|i-j|}}$ (here $S_0 = 0$). The sum S_n has the binomial distribution with parameters n, $p = \frac{2}{3}$, and $q = \frac{1}{3}$. Hence

$$\mathsf{E}(-1)^{S_n} = \sum_{k=0}^{n} (-1)^k \binom{n}{k} p^k q^{n-k} = (q-p)^n = (-\tfrac{1}{3})^n,$$

$$\mathsf{E}T_{k,k+m}^2 = \sum_{i,j=k}^{k+m} a_i a_j (-\tfrac{1}{3})^{|i-j|} = \sum_{i=k}^{k+m} a_i^2 + 2 \sum_{k\le i<j\le k+m} a_i a_j (-\tfrac{1}{3})^{j-i} \le$$

$$\le \sum_{i=k}^{k+m} a_i^2 + 2 \sum_{s=1}^{m} \sum_{k\le i\le k+m-s} |a_i a_{i+s}| \tfrac{1}{3^s}.$$

By the Cauchy–Schwarz inequality

$$\sum_{k\le i\le k+m-s} |a_i a_{i+s}| \le \left(\sum_{k\le i\le k+m-s} a_i^2 \right)^{1/2} \left(\sum_{k\le i\le k+m-s} a_{i+s}^2 \right)^{1/2} \le \sum_{i=k}^{\infty} a_i^2,$$

therefore,

$$\mathsf{E}T_{k,k+m}^2 \le \sum_{i=k}^{\infty} a_i^2 \left(1 + 2 \sum_{s=1}^{\infty} \tfrac{1}{3^s} \right) = 2 \sum_{i=k}^{\infty} a_i^2 \to 0, \text{ as } k \to \infty.$$

Thus, the partial sums of the series $\sum_{n=1}^{\infty} (-1)^{S_n} a_n$ indeed form a Cauchy sequence in $L_2(\Omega, \mathscr{F}, \mathsf{P})$, which proves the claim.

2014

1 For $\pi/2 \le x \le \pi$, it holds $\cos x \le 0 < e^{-x^2/2}$, hence it remains to consider $0 < x < \pi/2$. After taking the logarithm on both sides the inequality turns to $\ln \cos x < -x^2/2$, or equivalently $x^2/2 + \ln \cos x < 0$, $0 < x < \pi/2$. The function $f(x) = x^2/2 + \ln \cos x$ is decreasing on $[0, \pi/2)$, because $f'(x) = x - \tan x < 0$, $0 < x < \pi/2$. Therefore, $f(x) < f(0) = 0$, $0 < x < \pi/2$.

2 The value of initial polynomial $x^k(x-2)^{2n}$ at $x = 1$ equals 1. This property is preserved after all the allowed steps. Thus, it is impossible to get a polynomial of the form $x^l(x-2)^{2m+1}$, which equals -1 at $x = 1$.
Answer: no, it is impossible.

3 For arbitrary points P, Q, R, S in the space, define $f(P, Q, R, S) = \left(\overrightarrow{SP}, \overrightarrow{SQ}, \overrightarrow{SR}\right)$ (here $\left(\overrightarrow{a_1}, \overrightarrow{a_2}, \overrightarrow{a_3}\right) = \overrightarrow{a_1} \cdot \left(\overrightarrow{a_2} \times \overrightarrow{a_3}\right)$ is a scalar triple product). Then

$$V_{PQRS} = \tfrac{1}{6}|f(P, Q, R, S)|,$$

and it suffices to prove that $f(A_1, B_1, M_C, M_D) = f(A_2, B_2, M_C, M_D)$. Notice that

$$\begin{aligned} & f(A_1, B_1, M_C, M_D) + f(A_2, B_1, M_C, M_D) = \\ & = \left(\overrightarrow{M_DA_1} + \overrightarrow{M_DA_2}, \overrightarrow{M_DB_1}, \overrightarrow{M_DM_C}\right) = \left(2\overrightarrow{M_DM_A}, \overrightarrow{M_DM_B} + \overrightarrow{M_BB_1}, \overrightarrow{M_DM_C}\right) = \\ & = 2\left(\overrightarrow{M_DM_A}, \overrightarrow{M_DM_B}, \overrightarrow{M_DM_C}\right) = 2f(M_A, M_B, M_C, M_D). \end{aligned}$$

We used the equality $\left(\overrightarrow{M_DM_A}, \overrightarrow{M_BB_1}, \overrightarrow{M_DM_C}\right) = 0$, which is true because the vector $\overrightarrow{M_DB_1}$ lies in the plane ACD that is parallel to the plane $M_AM_CM_D$, hence it is a linear combination of vectors $\overrightarrow{M_DM_A}$ and $\overrightarrow{M_DM_C}$.

V. Brayman and A. Kukush, *Undergraduate Mathematics Competitions (1995–2016)*, Problem Books in Mathematics,
DOI 10.1007/978-3-319-58673-1_42

Similarly

$$f(A_2, B_2, M_C, M_D) + f(A_2, B_1, M_C, M_D) = 2f(M_A, M_B, M_C, M_D).$$

Subtract the obtained equalities to get the required equality.

4 Consider a polynomial $P(x) = x^4 + \beta x^3 - \alpha x + 1$. Choose $\alpha > 0$ and $\beta > 0$ which have the following properties:
1) for all $x > 0$, it holds $x^4 + 1 - \alpha x \geq 0$, and there exists a number $x_1 > 0$ such that $x_1^4 + 1 - \alpha x_1 = 0$, and
2) for all $x > 0$, it holds $x^4 + 1 + \beta x^3 \geq 0$, and there exists a number $x_2 < 0$ such that $x_2^4 + 1 + \beta x_2^3 = 0$.

It suffices to put $\alpha = \min\limits_{x>0} \frac{x^4+1}{x}$ and $\beta = -\max\limits_{x<0} \frac{x^4+1}{x^3}$. It is easy to check that the minimal and the maximal values exist and are attained at unique points $x_1 > 0$ and $x_2 < 0$, respectively.

The polynomial $P(x)$ has no real roots, because
1) if $x \geq 0$ then $P(x) = (x^4 + 1 - \alpha x) + \beta x^3 > 0$, and
2) if $x < 0$ then $P(x) = (x^4 + 1 + \beta x^3) - \alpha x > 0$.

It is evident that each of the polynomials $x^4 + \beta x^3 - \alpha x$ and $\beta x^3 - \alpha x + 1$ has a real root. Finally, each of the polynomials $x^4 - \alpha x + 1$ and $x^4 + \beta x^3 + 1$ has a real root due to the choice of α and β.
Answer: yes, there exists.

5 Consider $t_1, t_2 \in \left(0, \frac{\pi}{2}\right)$ such that $x_1 = \cos^2 t_1$ and $x_2 = \cos^2 t_2$. Then

$$\begin{aligned}\sqrt{x_1(1-x_2)} + \sqrt{x_2(1-x_1)} &= \cos t_1 \sin t_2 + \cos t_2 \sin t_1 = \sin(t_1 + t_2),\\ x_1 - x_2 &= \tfrac{1}{2}(\cos 2t_1 - \cos 2t_2) = \sin(t_1 + t_2)\sin(t_1 - t_2),\\ \max\left(|2x_1 - 1|, |2x_2 - 1|\right) &= \max\left(|\cos 2t_1|, |\cos 2t_2|\right),\end{aligned}$$

hence the inequality in question turns into

$$\sin(t_1 + t_2)\max\left(|\cos 2t_1|, |\cos 2t_2|\right) \geq |\sin(t_1 + t_2)\sin(t_1 - t_2)|.$$

Since $t_1 + t_2 \in (0, \pi)$, we have $\sin(t_1 + t_2) > 0$. Thus the inequality is equivalent to

$$\max\left(|\cos 2t_1|, |\cos 2t_2|\right) \geq |\sin(t_1 - t_2)|.$$

Because $2t_1, 2t_2 \in (0, \pi)$ and $(2t_1, 2t_2) \neq \left(\frac{\pi}{2}, \frac{\pi}{2}\right)$, there exists $s \in \left(0, \frac{\pi}{4}\right)$ such that

$$\max\left(|\cos 2t_1|, |\cos 2t_2|\right) = \cos 2s.$$

Then $t_1, t_2 \in \left[s, \frac{\pi}{2} - s\right]$, $|t_1 - t_2| \leq \frac{\pi}{2} - 2s$ and $|\sin(t_1 - t_2)| \leq \sin\left(\frac{\pi}{2} - 2s\right) = \cos 2s$, which finishes the proof.

The equality holds if and only if either $t_1 = s$ and $t_2 = \frac{\pi}{2} - s$ or $t_1 = \frac{\pi}{2} - s$ and $t_2 = s$ for some $s \in \left(0, \frac{\pi}{4}\right)$, i.e. $x_1 \in (0, \frac{1}{2}) \cup (\frac{1}{2}, 1)$ and $x_2 = 1 - x_1$.

6 Let P be $n \times n$ matrix. The set of vectors $H = \operatorname{Im} P = \{Px,\ x \in \mathbb{R}^n\}$ is a subspace of $\mathbb{R}^n$, and $H \neq \{0\}$, because P is not zero matrix. If $y = Px \in H$ then $Py = P^2x = Px = y$. Hence $H \neq \mathbb{R}^n$ because P is not the identity matrix. Show that there exists a matrix Q such that $Qx \in H$ for all $x \in \mathbb{R}^n$, $Qy = y$ for all $y \in H$ and $Q \neq P$.

Since P is neither zero matrix nor the identity one, there exist vectors $a, b \in \mathbb{R}^n$ such that $Pa \neq 0$ and $(I - P)^{\mathrm{T}}b \neq 0$. Then the matrix $Q = P + Pab^{\mathrm{T}}(I - P)$ has above-mentioned properties. Indeed, for all $x \in \mathbb{R}^n$ it holds $Qx = P(x + ab^{\mathrm{T}}(I - P)x) \in H$, and for all $y = Px \in H$ we have $Qy = Py + Pab^{\mathrm{T}}(I - P)Px = y + Pab^{\mathrm{T}}(P - P^2)x = y$. Finally $Q \neq P$, because $Q - P$ is a product of nonzero column vector Pa by nonzero row vector $b^{\mathrm{T}}(I - P) = ((I - P)^{\mathrm{T}}b)^{\mathrm{T}}$.

By the construction for all $x \in \mathbb{R}^n$ it holds $Px \in H$ and $Qx \in H$, hence $Q(Qx) = Qx$, $Q(Px) = Px$, and $P(Qx) = Qx$. Thus, $Q^2 = Q$, $QP = P$, and $PQ = Q$, whence $PQ = (QP)Q$ but $QP = P \neq Q = PQ$.

Answer: yes, always.

7 For $0 < \delta < \frac{1}{2}$ put $\alpha(\delta) = \max\limits_{x\in[0,1]} f(x) - \max\limits_{x\in A(\delta)} f(x)$, where

$$A(\delta) = \{x \in [0, 1] : |x - x_0| \geq \delta\}, \quad B(\delta) = \{x \in [0, 1] : |x - x_0| \leq \delta\}.$$

The condition of the problem implies that $\alpha(\delta) > 0$. Denote $\|g\| = \max\limits_{x\in[0,1]} |g(x)|$. If $\|g\| = 0$ then $g(x) = 0$, $x \in [0, 1]$, and the statement of the problem is evident. We assume further that $\|g\| > 0$.

Lemma 1 *For* $0 < t < \dfrac{\alpha(\delta)}{2\|g\|}$, *the maximum of the function* $f(\cdot) + t\,g(\cdot)$ *is attained on the segment* $[x_0 - \delta, x_0 + \delta]$.

Proof We have

$$\max_{x\in B(\delta)} (f(x) + t\,g(x)) - \max_{x\in A(\delta)} (f(x) + t\,g(x)) \geq$$
$$\geq \max_{x\in B(\delta)} f(x) - t\|g\| - \max_{x\in A(\delta)} f(x) - t\|g\| = \alpha(\delta) - 2t\|g\| > 0.$$

□

Fix an arbitrary $\varepsilon > 0$ and choose $\delta \in \left(0, \frac{1}{2}\right)$ such that $|g(x) - g(x_0)| \leq \varepsilon$ for all $x \in B(\delta)$. The lemma implies that for small enough $t > 0$, it holds

$$\frac{\varphi(t) - \varphi(0)}{t} = \frac{1}{t}\left(\max_{x\in[0,1]} (f(x) + t\,g(x)) - \max_{x\in[0,1]} f(x)\right) =$$
$$= \frac{1}{t}\left(\max_{x\in B(\delta)} (f(x) + t\,g(x)) - f(x_0)\right) \leq$$
$$\leq \frac{1}{t}\left(f(x_0) + t\max_{x\in B(\delta)} g(x) - f(x_0)\right) = \max_{x\in B(\delta)} g(x) \leq g(x_0) + \varepsilon.$$

Thus, $\limsup_{t\to 0+} \dfrac{\varphi(t)-\varphi(0)}{t} \le g(x_0)$.

On the other hand,

$$\frac{\varphi(t)-\varphi(0)}{t} = \frac{1}{t}\left(\max_{x\in[0,1]}(f(x)+t\,g(x)) - \max_{x\in[0,1]} f(x)\right) \ge$$

$$\ge \frac{1}{t}(f(x_0)+t\,g(x_0)-f(x_0)) = g(x_0).$$

Hence $\liminf_{t\to 0+} \dfrac{\varphi(t)-\varphi(0)}{t} \ge g(x_0)$. Therefore, $\varphi'_+(0) = \lim_{t\to 0+} \dfrac{\varphi(t)-\varphi(0)}{t} = g(x_0)$.

Notice that $\varphi(-t) = \max_{x\in[0,1]} (f(x)+t\cdot(-g(x)))$. Hence $\varphi'_-(0) = -(-g(x_0)) = g(x_0)$. Because $\varphi'_+(0) = \varphi'_-(0) = g(x_0)$, there exists $\varphi'(0) = g(x_0)$.

8 By substitution $t = \pi - x$ we get that the integral equals $\int_0^\pi (\sin t)^{-\cos t}\,dt$. Show that $(\sin t)^{-\cos t} \sim \frac{1}{t}$, as $t \to 0+$. By the limit comparison test this will imply that the integral diverges to $+\infty$. It holds

$$\frac{(\sin t)^{\cos t}}{t} \sim (\sin t)^{\cos t - 1} = e^{(\cos t-1)\ln \sin t}, \text{ as } t \to 0+.$$

Since $1 - \cos t = 2\sin^2 \frac{t}{2} \sim \frac{t^2}{2} \sim \frac{1}{2}\sin^2 t$, as $t \to 0+$, it holds that

$$(\cos t - 1)\ln \sin t \sim -\tfrac{1}{2}\sin^2 t \cdot \ln \sin t = \left| \begin{matrix} u = \sin t \\ u \to 0+ \end{matrix} \right| = -\tfrac{1}{2}u^2 \cdot \ln u \to 0, \text{ as } t \to 0+,$$

hence $e^{(\cos t-1)\ln \sin t} \to e^0 = 1$, as $t \to 0+$. Thus, $(\sin t)^{-\cos t} \sim \frac{1}{t}$, as $t \to 0+$, which finishes the proof.

Answer: $+\infty$.

9 The set $\mathscr{P}_{2014}$ of polynomials of degree at most 2014 is a finite-dimensional subspace of $L_{2014}([0, 2014], \lambda)$, where λ is Lebesgue measure on $\mathbb{R}$. In particular this set is closed. Hence the condition of the problem implies that the polynomial P also has degree at most 2014. Consider two norms on $\mathscr{P}_{2014}$: the L_{2014}-norm $\|f\|_{2014}$ and the norm $\|f\| = \|f(0)\| + \|f'\|_{2014}$. Every two norms on a finite-dimensional space are equivalent, hence the convergence of P_n to P in the first norm implies the convergence of P_n to P in the second norm, whence $\|P'_n - P'\|_{2014} \to 0$, as $n \to \infty$.

Answer: yes, it is.

10 If $f_n(x, y) \overset{\lambda_2}{\nrightarrow} 0$, as $n \to \infty$, then there exists $\varepsilon > 0$ such that λ_2-measure of the set $A_n = \{(x, y) \in [0, 1]^2 : |f_n(x, y)| \ge \varepsilon\}$ is not less than ε for infinitely many indices n.

By the property of product measure there exists $y_n \in [0, 1]$ such that for the cross section $A_{n,y_n} = \{x \in [0, 1] : (x, y_n) \in A_n\}$, it holds $\lambda_1(A_{n,y_n}) \ge \lambda_2(A_n)$. Consider $g_n(x) \equiv y_n$, $n \ge 1$. Then for all $x \in A_{n,y_n}$, it holds $|f_n(x, g_n(x))| = |f_n(x, y_n)| \ge \varepsilon$, and

$\lambda_1(A_{n,y_n}) \geq \lambda_2(A_n) \geq \varepsilon$ for infinitely many n. Thus, $f_n(x, g_n(x)) \overset{\lambda_1}{\nrightarrow} 0$, as $n \to \infty$, a contradiction.

13 For all $k \geq 4$, it holds

$$x_{k+1} = x_k - x_{k-1} + x_{k-2} + \varepsilon_{k+1} + \tfrac{1}{2}\varepsilon_k = \left(x_{k-1} - x_{k-2} + x_{k-3} + \varepsilon_k + \tfrac{1}{2}\varepsilon_{k-1}\right) - \\ - x_{k-1} + x_{k-2} + \varepsilon_{k+1} + \tfrac{1}{2}\varepsilon_k = x_{k-3} + \varepsilon_{k+1} + \tfrac{3}{2}\varepsilon_k + \tfrac{1}{2}\varepsilon_{k-1}.$$

Consider the sequences $y_k = x_{4k+1}$, $k \geq 0$, and $\xi_k = \varepsilon_{4k+1} + \frac{3}{2}\varepsilon_{4k} + \frac{1}{2}\varepsilon_{4k-1}$, $k \geq 1$. We have $y_k = y_{k-1} + \xi_k$, $k \geq 1$. Assume that $\mathsf{E}x_k^2 \leq C$ for all $k \geq 1$. Then it is evident that $\mathsf{E}y_k^2 \leq C$, $k \geq 0$. Since $\{\xi_k,\ k \geq 1\}$ is a sequence of independent random variables such that $\mathsf{E}\xi_k = 0$ and $\mathsf{Var}\xi_k = \frac{7}{2}$, $k \geq 1$, for random variables $y_k - y_0 = \sum_{i=1}^{k} \xi_i$, $k \geq 1$, it holds $\mathsf{E}(y_k - y_0) = 0$, and

$$\mathsf{E}(y_k - y_0)^2 = \mathsf{Var}(y_k - y_0) = \sum_{i=1}^{k} \mathsf{Var}\xi_i = \tfrac{7}{2}k \to \infty, \text{ as } k \to \infty.$$

On the other hand, $\mathsf{E}(y_k - y_0)^2 \leq 2\mathsf{E}y_0^2 + 2\mathsf{E}y_k^2 \leq 4C$, $k \geq 1$, a contradiction.
Answer: no, there is no such sequences.

14 Let P be $n \times n$ matrix and I be the identity $n \times n$ matrix. Then $I - P$ is neither zero matrix nor the identity one, and $(I - P)^2 = I - 2P + P^2 = I - P$. We apply the reasonings from solution of Problem 6 to the matrix $I - P$ instead of P, and get that there exists a matrix Q such that $Q^2 = Q$, $Q(I - P) = I - P$, $(I - P)Q = Q$, and $Q \neq I - P$. The matrix Q has required properties because it satisfies the relations $QP = Q + P - I$ and $PQ = 0$, whence $Q(PQ) = 0 = PQ$ and $(PQ)P = 0 \neq QP$.
Answer: yes, always.

2015

1 Let the endpoints of the segment have coordinates $(a, \frac{1}{a})$ and $(b, \frac{8}{b})$. Then the middle point has coordinates $(\frac{a+b}{2}, \frac{1}{2a} + \frac{4}{b})$. If this points lies on the hyperbola $y = \frac{1}{x}$ then $\frac{a+b}{2}(\frac{1}{2a} + \frac{4}{b}) = 1$, whence $\frac{1}{4} + \frac{b}{4a} + 2 + \frac{2a}{b} = 1$, $\frac{b}{4a} + \frac{2a}{b} = -\frac{5}{4}$, $(\frac{b}{a})^2 + 5\frac{b}{a} + 8 = 0$. But the quadratic equation $x^2 + 5x + 8 = 0$ has no real roots, a contradiction. *Answer:* no, it is impossible.

2 Introduce a polynomial $P(t) = (1 + tx_1)(1 + tx_2)\dots(1 + tx_n)$. It has a degree at most n. The system implies that $P(k) = k + 1$, $1 \le k \le n$, and moreover $P(0) = 1$. Hence the polynomial $P(t) - t - 1$ of degree at most n has $n + 1$ roots, therefore, it is identically zero. Thus, $(1 + tx_1)(1 + tx_2)\dots(1 + tx_n) \equiv 1 + t$. Hence one of the numbers $x_1, \dots, x_n$ equals 1 and the rest of them equal 0.
Answer: $(1, 0, \dots, 0)$ and all permutations of these numbers.

3 Assume that there exists a continuous and non-monotone function f, which satisfy $(f(z) - f(y))(f(y) - f(x)) \ge 0$ for all $0 \le x < y < z \le 1$ such that $z - y = y - x$. In particular for each $u, v \in [0, 1]$ if $f(u) \le f(v)$ then $f(u) \le f(\frac{u+v}{2}) \le f(v)$. Without loss of generality we may assume that $f(0) \le f(1)$ (otherwise consider the function $-f$ instead of f). Then we get subsequently

$$f(0) \le f(\tfrac{1}{2}) \le f(1),\ f(0) \le f(\tfrac{1}{4}) \le f(\tfrac{1}{2}) \le f(\tfrac{3}{4}) \le f(1),\ \dots$$

It is not difficult to prove by induction on n that

$$f(0) \le f(\tfrac{1}{2^n}) \le f(\tfrac{2}{2^n}) \le \dots \le f(\tfrac{2^n-1}{2^n}) \le f(1) \text{ for all } n \ge 0.$$

Show that the function f is nondecreasing on $[0, 1]$. For arbitrary numbers $0 \le a < b \le 1$, put $a_n = \frac{[2^n a]}{2^n}$, $b_n = \frac{[2^n b]}{2^n}$, $n \ge 0$. Since $[2^n a] \le [2^n b]$, it holds $f(a_n) \le f(b_n)$. Because $a_n \to a$ and $b_n \to b$, as $n \to \infty$, by continuity of f we get

V. Brayman and A. Kukush, *Undergraduate Mathematics Competitions (1995–2016)*, Problem Books in Mathematics,
DOI 10.1007/978-3-319-58673-1_43

$$f(a) = \lim_{n\to\infty} f(a_n) \le \lim_{n\to\infty} f(b_n) = f(b).$$

Thus, f is monotone on $[0, 1]$, and we come to a contradiction.

4 A bijection on the finite set X is a permutation of its elements. We decompose a permutation f into cycles. Consider a set $A \subset X$ such that $f(A) = A$. Then for each element $a \in A$ the set A contains the elements $f(a)$, $f(f(a))$, etc. Thus, the cycle which contains a is a subset of A. Hence A is a union of some cycles of the permutation f. On the other hand, if A is a union of some cycles of the permutation f then it is clear that $f(A) = A$. Therefore, if the permutation f consists of k cycles, then there exist exactly 2^k sets A for which $f(A) = A$.

5 If numbers n and k are odd, then Bilbo might have chosen either $x = 1$ and $y = -1$ or $x = 2$ and $y = -2$, and in both cases he tells Gollum the numbers 0 and 0. Similarly if numbers n and k are even, then Bilbo might have chosen either $x = 1$ and $y = 1$ or $x = 1$ and $y = -1$, and in both cases he tells Gollum the numbers 2 and 2. Therefore, if the numbers n and k have the same parity, then Gollum cannot determine xy.

From now on we assume that n is odd and k is even. Let Bilbo had chosen the numbers $x \le y$ and told Gollum the numbers $a = x^n + y^n$ and $b = x^k + y^k$. Then $x^n = \frac{a}{2} - t$, $y^n = \frac{a}{2} + t$ for some $t \ge 0$. Hence

$$x = \sqrt[n]{\tfrac{a}{2} - t}, \quad y = \sqrt[n]{\tfrac{a}{2} + t}, \quad \text{and} \quad x^k + y^k = \sqrt[n]{\left(\tfrac{a}{2} - t\right)^k} + \sqrt[n]{\left(\tfrac{a}{2} + t\right)^k}.$$

Since $xy = \sqrt[n]{\frac{a^2}{4} - t^2}$ is a decreasing function of t for $t \ge 0$, Gollum can determine xy if and only if there is no b such that the equality

$$\sqrt[n]{\left(\tfrac{a}{2} - t\right)^k} + \sqrt[n]{\left(\tfrac{a}{2} + t\right)^k} = b$$

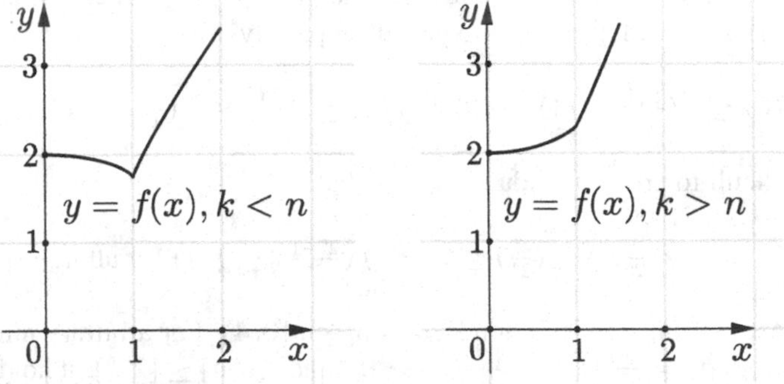

Fig. 1 The graph of the function $f(x) = \sqrt[n]{(1-x)^k} + \sqrt[n]{(1+x)^k}$

holds for two distinct values $t \geq 0$. It is evident that the latter condition holds if $a = 0$. For $a \neq 0$, after the change of variables $t = \frac{|a|}{2}s$, $b = \sqrt[n]{(\frac{a}{2})^k}\, c$ the equality takes a form $\sqrt[n]{(1-s)^k} + \sqrt[n]{(1+s)^k} = c$. Thus, Gollum can determine xy, if the function $f(s) = \sqrt[n]{(1-s)^k} + \sqrt[n]{(1+s)^k}$ attains each value at most once on $s \geq 0$. Determining intervals of monotonicity we get that for $k/n < 1$ the continuous function $f(s)$ decreases on $(0, 1)$ and increases on $(1, +\infty)$, and for $k/n > 1$, it increases on $[0, +\infty)$ (see Fig. 1). Thus, Gollum wins if and only if $k > n$.
Answer: either n is odd, k is even and $n < k$, or k is odd, n is even and $k < n$.

6 Introduce a function

$$f(x) = \frac{\sum\limits_{i=0}^{n} a_i e^{a_i x}}{\sum\limits_{i=0}^{n} e^{a_i x}}, \quad x \in \mathbb{R}.$$

Its derivative

$$f'(x) = \frac{\sum\limits_{i=0}^{n} a_i^2 e^{a_i x} \cdot \sum\limits_{i=0}^{n} e^{a_i x} - \left(\sum\limits_{i=0}^{n} a_i e^{a_i x}\right)^2}{\left(\sum\limits_{i=0}^{n} e^{a_i x}\right)^2}$$

is positive for every $x \in \mathbb{R}$. Indeed, by the Cauchy–Schwarz inequality the numerator is nonnegative and turns to 0 only if the vectors

$$\left(a_0 e^{a_0 x/2}, \ldots, a_n e^{a_n x/2}\right) \text{ and } \left(e^{a_0 x/2}, \ldots, e^{a_n x/2}\right)$$

are proportional. But this condition fails because not all the numbers $a_0, \ldots, a_n$ coincide. Hence the function f is strictly increasing on $\mathbb{R}$.

Put $m = \min\{a_i, 0 \leq i \leq n\}$ and $M = \max\{a_i, 0 \leq i \leq n\}$. Then $f(x) \to m$, as $x \to -\infty$, and $f(x) \to M$, as $x \to +\infty$. Therefore, a strictly monotone continuous function f attains each value from the interval (m, M) exactly once. It remains to notice that the equation from the statement of the problem can be written in a form $f(x) = \frac{1}{2^n} \sum\limits_{i=0}^{n} \binom{n}{i} a_i$, and $m < \frac{1}{2^n} \sum\limits_{i=0}^{n} \binom{n}{i} a_i < M$, because not all the numbers $a_0, \ldots, a_n$ coincide.

7 First we show that $\det \begin{pmatrix} A & B \\ C^{\mathrm{T}} & 0 \end{pmatrix} = \det \begin{pmatrix} A & B \\ C^{\mathrm{T}} & 1 \end{pmatrix}$. Indeed, in the decomposition of the determinant of the matrix $\begin{pmatrix} A & B \\ C^{\mathrm{T}} & d \end{pmatrix}$ by the last row, the number d is multiplied by $\det A = 0$, hence the determinant does not depend on the choice of d.

Denote by I the identity $n \times n$ matrix and by O the zero $n \times 1$ vector. Then

$$\begin{pmatrix} A & B \\ C^{\mathrm{T}} & 1 \end{pmatrix} = \begin{pmatrix} I & B \\ O^{\mathrm{T}} & 1 \end{pmatrix} \begin{pmatrix} A - BC^{\mathrm{T}} & O \\ C^{\mathrm{T}} & 1 \end{pmatrix},$$

hence

$$\det \begin{pmatrix} A & B \\ C^{\mathrm{T}} & 0 \end{pmatrix} = \det \begin{pmatrix} A & B \\ C^{\mathrm{T}} & 1 \end{pmatrix} = \det \begin{pmatrix} I & B \\ O^{\mathrm{T}} & 1 \end{pmatrix} \det \begin{pmatrix} A - BC^{\mathrm{T}} & O \\ C^{\mathrm{T}} & 1 \end{pmatrix} = 1 \cdot \det(A - BC^{\mathrm{T}}).$$

This implies the statement of the problem.

8 Since $f_n \to 0 \pmod{\lambda_1}$, as $n \to \infty$ (where λ_1 is Lebesgue measure on $\mathbb{R}$), it suffices to check whether the sequence converges to 0 in Lebesgue measure. We prove that there is no convergence in the measure. To this purpose wc show that

$$\begin{aligned} \lambda_1 \left(\left\{ x \in \mathbb{R} : |f_n(x)| \geq \tfrac{1}{2} \right\} \right) =& \lambda_1 \left(\left\{ x \in [0, n] : |\sin \pi x| \geq \tfrac{1}{\sqrt[n]{2}} \right\} \right) = \\ =& n \left(1 - \tfrac{2}{\pi} \arcsin \tfrac{1}{\sqrt[n]{2}} \right) \not\to 0. \end{aligned}$$

Indeed, for $t \in [0, 1]$, it holds $\arcsin t \leq \frac{\pi}{2} t$. Hence

$$n \left(1 - \tfrac{2}{\pi} \arcsin \tfrac{1}{\sqrt[n]{2}} \right) \geq n \left(1 - \tfrac{1}{\sqrt[n]{2}} \right) = \frac{2^{1/n} - 1}{2^{1/n} \cdot 1/n} \to \ln 2 > 0, \text{ as } n \to \infty.$$

Answer: the sequence does not converge in the measure.

9 Since $B \subset f(\mathbb{R}^2) \subset \mathbb{R}$ is a compact set, the numbers $y_* = \min B$ and $y^* = \max B$ are well defined, moreover $y_* = f(x_*)$ and $y^* = f(x^*)$ for some $x_*, x^* \in \mathbb{R}^2$. Consider

$$[x_*, x^*] = \{(1-t)x_* + tx^*,\ t \in [0, 1]\} \subset \mathbb{R}^2, \quad A = [x_*, x^*] \cap f^{-1}(B).$$

Then $[x_*, x^*]$ is a compact set, the set $f^{-1}(B)$ is closed in $\mathbb{R}^2$ as inverse image of a closed set under a continuous mapping, and the set A is compact as a closed subset of a compact set. It remains to show that $f(A) = B$. Introduce a function

$$g(t) = f((1-t)x_* + tx^*),\ t \in [0, 1].$$

Then $g(0) = f(x_*) = y_*$, $g(1) = f(x^*) = y^*$ and $g \in C([0, 1])$. Hence g attains all the values from $[y_*, y^*]$ on the segment $[0, 1]$. Thus, f attains all the values from $[y_*, y^*]$ on $[x_*, x^*]$. Therefore, $B \subset f([x_*, x^*])$, whence $f(A) = f([x_*, x^*] \cap f^{-1}(B)) = B$.

10 Put $m = \mathsf{E}\xi$ and $\delta = \frac{1}{3C}$. Consider an arbitrary random variable $\xi \in K_C$. Since

$$\mathsf{P}(|\xi - m| < \delta) = \int_{m-\delta}^{m+\delta} p(x)dx \le 2C\delta = \tfrac{2}{3},$$

it holds $\mathsf{P}(|\xi - m| \ge \delta) \ge \frac{1}{3}$. Then

$$\mathsf{Var}\,\xi = \mathsf{E}(\xi - m)^2 \ge \mathsf{E}\,(\xi - m)^2\,\mathrm{I}_{\{|\xi - m|\ge\delta\}} \ge \delta^2\,\mathsf{P}\,(|\xi - m| \ge \delta) \ge \tfrac{\delta^2}{3} = \tfrac{1}{27C^2}.$$

Thus, one can take $a(C) = \frac{1}{27C^2}$.

12 Introduce a function $f(a) = \frac{\mathsf{E}e^{2a\xi}}{\mathsf{E}e^{a\xi}}$, $a \ge 0$. Show that the function is continuous. For $0 \le a \le c$, it holds

$$e^{a\xi} \le e^{c\xi}\,\mathrm{I}_{\{\xi\ge 0\}} + \mathrm{I}_{\{\xi<0\}} \le e^{c\xi} + 1 \in L(\Omega, \mathcal{F}, \mathsf{P}).$$

Hence the function $g(a) = \mathsf{E}e^{a\xi}$ is continuous on $[0, c]$ for every $c > 0$. Because $g(a) > 0$ for $a \ge 0$, the function $f(a) = \frac{g(2a)}{g(a)}$ is continuous as well.

Show that $f(a) \to +\infty$, as $a \to +\infty$. Notice that by the continuity of measure from below it holds $\lim\limits_{n\to\infty} \mathsf{P}(\xi > \frac{1}{n}) = \mathsf{P}(\xi > 0) > 0$, hence for some $N \ge 1$ it holds $\mathsf{P}(\xi > \frac{1}{N}) > 0$. Then

$$f(a) = \frac{\mathsf{E}e^{2a\xi}}{\mathsf{E}e^{a\xi}} \ge \mathsf{E}e^{a\xi} \ge e^{\frac{a}{N}}\,\mathsf{P}(\xi > \tfrac{1}{N}) \to +\infty,\ a \to +\infty.$$

Thus, $f \in C([0, +\infty)$, $f(0) = 1$, and $f(a) \to +\infty$, as $a \to +\infty$. Therefore, by the intermediate value theorem there exists a number $\sigma > 0$ such that $f(\sigma) = 2$.

14 We parametrize the equation:

$$\begin{cases} y = xu - \sqrt{1+u^2}, \\ dy = udx. \end{cases}$$

Then for the variables u and x, we have $\left(x - \frac{u}{\sqrt{1+u^2}}\right) du = 0$, whence

$$\begin{cases} u = c,\ c \in \mathbb{R}, \\ y = cx - \sqrt{1+c^2} \end{cases} \quad\text{or}\quad \begin{cases} x = \frac{u}{\sqrt{1+u^2}}, \\ y = \frac{u^2}{\sqrt{1+u^2}} - \sqrt{1+u^2} = -\frac{1}{\sqrt{1+u^2}}, \end{cases}$$

i.e., $y = cx - \sqrt{1+c^2}$, $c \in \mathbb{R}$, or $y^2 + x^2 = 1$, $y < 0$.

One can draw tangent lines from the points (x_1, y_1) and (x_2, y_2) to the lower half of the unit circle, besides the points of tangency will belong to the third and the fourth quadrants, respectively. It is easy to verify that the equation of each tangent line to the lower half of the unit circle has a form $y = cx - \sqrt{1+c^2}$ for some $c \in \mathbb{R}$. Hence the graph of a required solution can be constructed as a union of two segments of tangent lines and an arc of the unit circle as shown on Fig. 2.

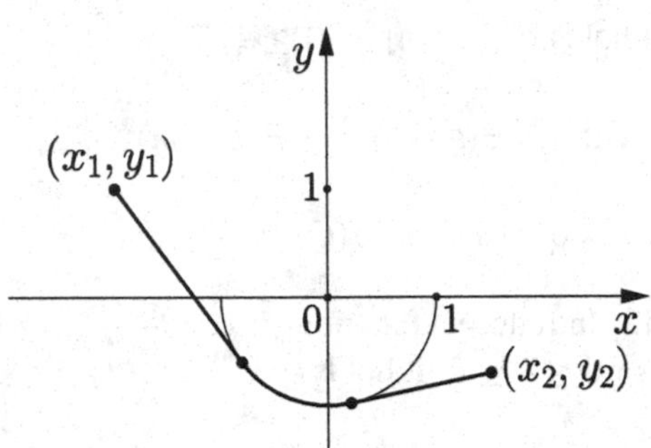

Fig. 2 The graph of a solution to the differential equation which connects the points (x_1, y_1) and (x_2, y_2)

2016

1 If n is divisible by 9, then $\cos\frac{n\pi}{9} = \pm 1$ and the expression equals $4 - \sqrt[3]{5}$. If n is divisible by 3 and is not divisible by 9, then $\cos\frac{n\pi}{9} = \pm\frac{1}{2}$ and the expression equals $1 + \sqrt[3]{4}$. Let n be not divisible by 3. Rewrite the expression as

$$2 + 2\cos\frac{2n\pi}{9} + \sqrt[3]{1 - 6\cos\frac{2n\pi}{9}}.$$

Since $\cos 3\alpha = 4\cos^3\alpha - 3\cos\alpha$, it holds for $\alpha = \frac{2n\pi}{9}$ that

$$4\cos^3\frac{2n\pi}{9} = \cos\frac{6n\pi}{9} + 3\cos\frac{2n\pi}{9} = -\frac{1}{2} + 3\cos\frac{2n\pi}{9},$$

whence

$$2 + 2\cos\frac{2n\pi}{9} + \sqrt[3]{1 - 6\cos\frac{2n\pi}{9}} = 2 + 2\cos\frac{2n\pi}{9} + \sqrt[3]{-8\cos^3\frac{2n\pi}{9}} = 2.$$

It remains to notice that $4 - \sqrt[3]{5} > 2$ and $1 + \sqrt[3]{4} > 2$.

Answer: 2.

2 For each infinite increasing geometric progression which consists of positive integers its first term and the denominator are positive integers as well. Hence the set of all such progressions is countable. We enumerate all such progressions and choose a number $c_n > 10^n$ from the nth progression. Introduce a set $A = \mathbb{N} \setminus \{c_1, c_2, \ldots\} = \{a_1, a_2, \ldots\}$. It does not contain any infinite increasing geometric progression, and

$$\frac{\max\{k : a_k \le n\}}{n} \ge \frac{n - [\lg n]}{n} \to 1, \text{ as } n \to \infty.$$

V. Brayman and A. Kukush, *Undergraduate Mathematics Competitions (1995–2016)*, Problem Books in Mathematics,
DOI 10.1007/978-3-319-58673-1_44

Alternative solution. Denote

$$B = \{bq^{b+1},\ b, q \in \mathbb{N},\ q \geq 2\} \text{ and } A = \mathbb{N} \setminus B = \{a_1, a_2, \ldots\}.$$

Each infinite increasing geometric progression which consists of positive integers has a positive integer first term b and a positive integer denominator $q \geq 2$, hence it contains the number bq^{b+1}. Therefore, the set A contains no infinite increasing geometric progressions. Estimate a number of elements of B which do not exceed a positive integer n. If $bq^{b+1} \leq n$ then $2^b < bq^{b+1} \leq n$, whence $b < \log_2 n$ and $q^2 \leq bq^{b+1} \leq n$, whence $q \leq \sqrt{n}$. Thus, the set B contains less than $\sqrt{n}\log_2 n$ numbers, so the set A contains more than $n - \sqrt{n}\log_2 n$ numbers that do not exceed n. Therefore,

$$\frac{\max\{k : a_k \leq n\}}{n} > \frac{n - \sqrt{n}\log_2 n}{n} \to 1, \text{ as } n \to \infty.$$

Answer: yes, there exists.

3 Without loss of generality we may assume that $\angle B_1A_1C_1 = 90°$ and either $\angle BAC = 90°$ or $\angle ABC = 90°$. Introduce a Cartesian coordinate system in such a way that $\overrightarrow{A_1B_1} = (1, 0, 0)$ and $\overrightarrow{A_1C_1} = (0, 1, 0)$. Then for some $x, y \in \mathbb{R}$, it holds $\overrightarrow{AB} = (1, 0, x)$, $\overrightarrow{AC} = (0, 1, y)$, and $\overrightarrow{BC} = (-1, 1, y - x)$.

Let $\angle BAC = 90°$. Then the triangle ABC is a right isosceles triangle if $\left|\overrightarrow{AB}\right| = \left|\overrightarrow{AC}\right|$ and $\left(\overrightarrow{AB}, \overrightarrow{AC}\right) = 0$, i.e., $1 + x^2 = 1 + y^2$ and $xy = 0$. Thus, $x = y = 0$ and $AB = A_1B_1$.

Let $\angle ABC = 90°$. Then the triangle ABC is a right isosceles triangle if $\left|\overrightarrow{AB}\right| = \left|\overrightarrow{BC}\right|$ and $\left(\overrightarrow{AB}, \overrightarrow{BC}\right) = 0$, i.e.,

$$\begin{cases} 1 + x^2 = 2 + (y - x)^2, \\ -1 + x(y - x) = 0, \end{cases} \qquad \begin{cases} y^2 - 2xy = -1, \\ xy - x^2 = 1. \end{cases}$$

Since $x \neq 0$ and $y^2 - xy - x^2 = 0$, it holds $(y/x)^2 - y/x - 1 = 0$. Hence $y = \frac{1}{2}\left(1 \pm \sqrt{5}\right)x$. Denote $\varphi = \frac{1}{2}\left(1 + \sqrt{5}\right)$. For $y = \frac{1}{2}\left(1 - \sqrt{5}\right)x$ the system of equations has no solution, and for $y = \varphi x$ it is easy to find solutions of the system: $x = \sqrt{\varphi}$, $y = \varphi\sqrt{\varphi}$ and $x = -\sqrt{\varphi}$, $y = -\varphi\sqrt{\varphi}$. Thus, $AB = \sqrt{1 + x^2} = \sqrt{1 + \varphi} = \varphi = \varphi A_1B_1$.

Answer: 1 and $\frac{1+\sqrt{5}}{2}$.

4 Put $y_n = \frac{1}{x_n}$. Then $y_0 = 1$, and for $n \geq 0$, it holds

$$\frac{1}{y_{n+1}} = \frac{1}{y_n} - \frac{1}{2016y_n^2}, \quad \text{or equivalently} \quad y_{n+1} = \frac{2016y_n^2}{2016y_n - 1},$$

whence

$$y_{n+1} - y_n = \frac{1}{2016 - \frac{1}{y_n}}.$$

It is clear that $x_n \in (0, 1)$ for all $n \geq 1$. Hence $y_n > 1$ for all $n \geq 1$. Thus,

$$y_1 - y_0 = \frac{1}{2015} \quad \text{and} \quad \frac{1}{2016} < y_{n+1} - y_n = \frac{y_n}{2016 y_n - 1} < \frac{1}{2015}, \quad n \geq 1.$$

By induction on n we get

$$1 + \frac{n}{2016} < y_n < 1 + \frac{n}{2015}, \quad n \geq 2.$$

Therefore, $y_{2015} < 2 < y_{2016}$, that is $x_{2016} < \frac{1}{2} < x_{2015}$.

5 (a) Let A be the identity matrix and B be a block-diagonal matrix with 1008 blocks of the form $\left(\begin{smallmatrix} 0 & -1 \\ 1 & 0 \end{smallmatrix}\right)$ on the diagonal. Then $aA + bB$ is a block-diagonal matrix with 1008 blocks of the form $\left(\begin{smallmatrix} a & -b \\ b & a \end{smallmatrix}\right)$ on the diagonal. Therefore, $\det(aA + bB) = (a^2 + b^2)^{1008} \neq 0$ if $a^2 + b^2 \neq 0$.

(b) If A or B is singular, one can put $a = 1, b = 0$ or $a = 0, b = 1$, respectively. Let both matrices A and B be nonsingular. Since $\det(-A) = (-1)^{2017} \det A = -\det A$, there exists a number $\alpha \in \{-1; 1\}$ such that $\det(\alpha A) > 0$. Similarly there exists a number $\beta \in \{-1; 1\}$ such that $\det(\beta B) < 0$. Then the function

$$f(t) = \det\left(t\alpha A + (1 - t)\beta B\right), \quad t \in \mathbb{R},$$

is continuous, $f(0) = \det(\beta B) < 0$, and $f(1) = \det(\alpha A) > 0$. Thus, by the intermediate value theorem there exists a real number $0 < t_0 < 1$, for which $f(t_0) = 0$, and one can put $a = t_0\alpha$, $b = (1 - t_0)\beta$.
Answer: (a) no, (b) yes.

6 Prove that

$$\sqrt{x} > \frac{m}{n} + \frac{1}{mn} = \frac{m^2 + 1}{mn}.$$

Since $\sqrt{x} > \dfrac{m}{n}$, it holds $n^2 x > m^2$. But $n^2 x \neq m^2 + 1$ and $n^2 x \neq m^2 + 2$, because the numbers $m^2 + 1$ and $m^2 + 2$ are not divisible by 7 for all integer m. Hence $n^2 x \geq m^2 + 3$.

Suppose that $\sqrt{x} < \dfrac{m^2 + 1}{mn}$. Then $m^2 n^2 x \leq (m^2 + 1)^2 = m^4 + 2m^2 + 1$, hence $m^4 + 2m^2 + 1 \geq m^2(m^2 + 3) = m^4 + 3m^2$, i.e., $1 \geq m^2$, whence $m = 1$. But then $n^2 x \leq 4$ and x is not divisible by 7, a contradiction.

Therefore, if $\sqrt{x} > \frac{m}{n}$ then $\sqrt{x} > \frac{m}{n} + \frac{1}{mn} = \frac{m^2+1}{mn}$. This implies

$$\sqrt{x} > \frac{m^2+1}{mn} + \frac{1}{(m^2+1)mn} = \frac{m^4+2m^2+2}{m^3n+mn}.$$

7 Introduce a polynomial

$$P(y) = (x_1+y)(x_2+y)\dots(x_n+y) - c.$$

It holds $P(y_1) = P(y_2) = \dots = P(y_n) = 0$, besides P is a polynomial of degree n with leading coefficient 1. Therefore,

$$P(y) = (y-y_1)(y-y_2)\dots(y-y_n) = (x_1+y)(x_2+y)\dots(x_n+y) - c.$$

Hence

$$P(-y) = (x_1-y)(x_2-y)\dots(x_n-y) - c = (-y-y_1)(-y-y_2)\dots(-y-y_n),$$

thus,

$$(y+y_1)(y+y_2)\dots(y+y_n) = (y-x_1)(y-x_2)\dots(y-x_n) - (-1)^n c.$$

Substitute $y = x_i$ and get $(x_i+y_1)(x_i+y_2)\dots(x_i+y_n) = (-1)^{n+1}c$.

Answer: $(-1)^{n+1}c$.

8 Let the numbers written on the opposite faces of the die be a_1 and a_2, b_1 and b_2, c_1 and (c_2), respectively. If the cubic die satisfies the condition of the problem, then the sum of the numbers written on each pair of adjacent faces is integer. A total sum of the written numbers equals $S = (a_1+b_1)+(b_2+c_1)+(c_2+a_2)$, hence the sum is integer as well. But expectation of the sum of numbers on the two upper faces equals $\frac{S}{3}$. Hence it should hold $\frac{S}{3} = \frac{21}{6}$, i.e., $S = 10.5$. Thus, S is not integer, we get a contradiction.

Answer: no, it is impossible.

9 The sets $A = \{x \in [0,1] : f(x) = 0\}$ and $B = \{x \in [0,1] : f(x) = f'(x) = 0\}$ are closed, hence Borel measurable. It is clear that $B \subset A$. Show that all the limit points of A belong to B. Indeed, assume that $\{x_n,\ n \ge 1\} \subset A \setminus \{x\}$ and $x_n \to x$, as $n \to \infty$. Then $x \in A$, because the set A is closed. Hence $f(x_n) = 0$, $n \ge 1$, and $f(x) = 0$, whence $f'(x) = \lim\limits_{n\to\infty} \frac{f(x_n)-f(x)}{x_n-x} = 0$. Thus, $x \in B$. Therefore, the set $A \setminus B$ contains only isolated points of A. Then $A \setminus B$ is at most countable, because for each $n \ge 1$, there exists finitely many points of the set A, in $\frac{1}{n}$-neighborhood of which there is no other point of A. Therefore, $\lambda_1(A \setminus B) = 0$, and $\lambda_1(A) = \lambda_1(B) + \lambda_1(A \setminus B) = \lambda_1(B)$.

11 Since the matrices $A(t)$ are skew-symmetric, that is $A(t)^{\mathrm{T}} = -A(t), t \in \mathbb{R}$, it holds that

$$\frac{d}{dt}(X(t)x, X(t)x) = (A(t)X(t)x, X(t)x) + (X(t)x, A(t)X(t)x) =$$
$$= (A(t)X(t)x, X(t)x) + \left(A(t)^{\mathrm{T}}X(t)x, X(t)x\right) = 0$$

for all $x \in \mathbb{R}^n$ and $t \in \mathbb{R}$. Whence $\|X(t)x\|^2 = (X(t)x, X(t)x) = (x, x) = \|x\|^2$, i.e., the mapping $X(t) : \mathbb{R}^n \to \mathbb{R}^n$ is an isometry. Thus, the mapping $X(t) : S_r(0) \to S_r(0)$ is a bijection for each $t \in \mathbb{R}$ and each $r > 0$, where $S_r(0)$ is a sphere in $\mathbb{R}^n$ with center in the origin and radius r.

Consider a sequence of points $y_t = X(t)^{-1}y,\ t \in \mathbb{N}$. Let $r = \|y\|$. Since $\{y_t,\ t \geq 1\} \subset S_r(0)$ and $S_r(0)$ is a compact set, there exists a subsequence $\{y_{t_i},\ i \geq 1\}$ which converges to some $z \in S_r(0)$. Then

$$\|X(t_i)z - y\| = \|X(t_i)z - X(t_i)X(t_i)^{-1}y\| = \|z - y_{t_i}\| \to 0, \text{ as } i \to \infty.$$

12 For all $a, b \in \mathbb{R}$, it holds

$$\int_0^1 p(x, a, b)dx = \exp(f(a, b)) \cdot \int_0^1 \exp\left(ax + bx^2 + g(x)\right)dx = 1$$

(hereafter all the integrals are proper Riemann integrals). Hence

$$f(a, b) = -\ln\left(\int_0^1 \exp\left(ax + bx^2 + g(x)\right)dx\right),$$

therefore, $f \in C^{(1)}(\mathbb{R}^2)$. We differentiate the identity $\int_0^1 p(x, a, b)dx = 1$ with respect to a and b. We get

$$\int_0^1 \left(x + f_a'(a, b)\right)p(x, a, b)dx = 0, \quad \int_0^1 \left(x^2 + f_b'(a, b)\right)p(x, a, b)dx = 0.$$

Let $\xi = \xi(a, b)$ be a random variable with probability density function $p(x, a, b)$. Then

$$\mathsf{E}\xi = \int_0^1 xp(x, a, b)dx = -\int_0^1 f_a'(a, b)p(x, a, b)dx = -f_a'(a, b),$$
$$\mathsf{E}\xi^2 = \int_0^1 x^2p(x, a, b)dx = -\int_0^1 f_b'(a, b)p(x, a, b)dx = -f_b'(a, b),$$

whence $\mathsf{Var}\xi = \mathsf{E}\xi^2 - (\mathsf{E}\xi)^2 = -f_b'(a, b) - (-f_a'(a, b))^2 > 0$, i.e., $(f_a'(a, b))^2 + f_b'(a, b) < 0$ (notice that $\mathsf{Var}\xi > 0$, because if $\mathsf{Var}\xi = 0$, the random variable ξ would have no probability density).

13 Show that the limit in question equals zero. Since $K\left(1-\frac{1}{3^k}\right)=1-\frac{1}{2^k}$, it holds

$$K(x)\le 1-\frac{1}{2^k},\quad \text{for } 1-\frac{1}{3^{k-1}}\le x\le 1-\frac{1}{3^k},\quad k\ge 1.$$

Hence

$$I_n := n\int_{[0,1]} K^n(x)\,d\lambda_1 \le \sum_{k\ge 1} n\left(1-\frac{1}{2^k}\right)^n \frac{2}{3^k}.$$

For a fixed k, the sequence $a_n = n\left(1-\frac{1}{2^k}\right)^n$, $n\ge 1$, satisfies

$$\frac{a_{n+1}}{a_n} = \frac{n+1}{n}\left(1-\frac{1}{2^k}\right)\ge 1 \Leftrightarrow n\le 2^k-1,$$

hence the sequence is not decreasing for $n\le 2^k-1$ and not increasing for $n\ge 2^k$. Thus, $a_n\le a_{2^k}<2^k$, for all $n\in\mathbb{N}$.

For each k_0 it holds

$$I_n \le \sum_{1\le k\le k_0} n\left(1-\frac{1}{2^k}\right)^n\frac{2}{3^k} + \sum_{k>k_0} n\left(1-\frac{1}{2^k}\right)^n\frac{2}{3^k} <$$
$$< \sum_{1\le k\le k_0} n\left(1-\frac{1}{2^k}\right)^n\frac{2}{3^k} + \sum_{k>k_0}\frac{2^{k+1}}{3^k}.$$

For each $\varepsilon>0$, one can choose a number k_0 such that the second sum is less than $\varepsilon/2$. Since for each fixed k, it holds $n\left(1-\frac{1}{2^k}\right)^n\to 0$, as $n\to\infty$, for given k_0 and for all n large enough the first sum is less than $\varepsilon/2$ as well.

Alternative solution. For each $j\in\mathbb{N}$, introduce a function

$$K_j(x)=\begin{cases}1-\dfrac{1}{2^j}, & 0\le x\le 1-\dfrac{2}{3^j},\\[2mm] 1+\dfrac{3^j}{2^{j+1}}(x-1), & 1-\dfrac{2}{3^j}\le x\le 1.\end{cases}$$

It is not difficult to verify that $K(x)\le K_j(x)$, $x\in[0,1]$, and

$$n\int_{[0,1]} K_j^n(x)\,d\lambda_1 = n\left(1-\frac{2}{3^j}\right)\left(1-\frac{1}{2^j}\right)^n +$$
$$+\frac{n}{n+1}\frac{2^{j+1}}{3^j}\left(1-\left(1-\frac{1}{2^j}\right)^{n+1}\right)\to\frac{2^{j+1}}{3^j},\quad \text{as } n\to\infty.$$

Thus,

$$\limsup_{n\to\infty} n \int_{[0,1]} K^n(x)\,d\lambda_1 \le \lim_{n\to\infty} n \int_{[0,1]} K_j^n(x)\,d\lambda_1 = \frac{2^{j+1}}{3^j}.$$

Because $j \ge 1$ is arbitrary, the upper limit equals 0.

Answer: 0.

14 Find a matrix X_0 of the form $X_0 = A^{\mathrm{T}} Y$ for which $AX_0 = I$ (here Y is an $m \times m$ matrix). It holds $AA^{\mathrm{T}} Y = I$. Notice that AA^{T} is the Gram matrix of the rows of A, and it is nonsingular because $\operatorname{rk} A = m$. Hence $Y = (AA^{\mathrm{T}})^{-1}$ and $X_0 = A^{\mathrm{T}}(AA^{\mathrm{T}})^{-1}$.

Show that X_0 is a required matrix. If X is an arbitrary solution of the equation $AX = I$, then $A(X - X_0) = O$, where O is zero matrix of size $m \times m$. Then $(X - X_0)^{\mathrm{T}} A^{\mathrm{T}} = O$. Therefore, $(X - X_0)^{\mathrm{T}} X_0 = (X - X_0)^{\mathrm{T}} A^{\mathrm{T}}(AA^{\mathrm{T}})^{-1} = O$. Thus, the columns of the matrix $X - X_0$ are orthogonal to the corresponding columns of X_0. Thus, the sum of squared norms of the columns of $X = X_0 + (X - X_0)$ is not less than the sum of squared norms of the columns of X_0. The equality is attained at $X = X_0$ only.

Answer: $X = A^{\mathrm{T}}(AA^{\mathrm{T}})^{-1}$.

References

1. Alexanderson, G.L., Klosinski, L.F., Larson, L.C. (eds.): The William Lowell Putnam Mathematical Competition Problems and Solutions: 1965–1984. Math. Association of America, Washington (1985)
2. Dorogovtsev, A.Ya., Kukush, A.G.: Selected problems of university round of Mathematical Olympiad "A Student and Scientific-Technical Progress" [in Russian]. In: Mathematics Today, ed. by A.Ya. Dorogovtsev, pp. 125–165. Vyshcha Shkola, Kiev (1983)
3. Gleason, A.M., Greenwood, R.E., Kelly, L.M. (eds.): The William Lowell Putnam Mathematical Competition Problems and Solutions: 1938–1964. Math. Association of America, Washington (1980)
4. Kedlaya, K.S., Poonen, B., Vakil, R. (eds.): The William Lowell Putnam Mathematical Competition 1985–2000: Problems, Solutions, and Commentary. Math. Association of America, Washington (2002)
5. Konstantinov, O.Yu., Kukush, A.G.: Mathematical Olympiad in Kiev University, 1990 [in Russian]. In: Mathematics Today '93, ed. by A.Ya. Dorogovtsev, pp. 195–213. Vyshcha Shkola, Kiev (1993)
6. Kukush, A.G.: Problems of Mathematical Olympiad "A Student and Scientific-Technical Progress" [in Russian]. In: Mathematics Today '86, ed. by A.Ya. Dorogovtsev, pp. 160–193. Vyshcha Shkola, Kiev (1986)

V. Brayman and A. Kukush, *Undergraduate Mathematics Competitions (1995–2016)*, Problem Books in Mathematics,
DOI 10.1007/978-3-319-58673-1

Thematic Index

V. Brayman and A. Kukush, *Undergraduate Mathematics Competitions (1995–2016)*, Problem Books in Mathematics, DOI 10.1007/978-3-319-58673-1

Printed in the United States
By Bookmasters